D0857517

INTERNATIONAL POLITICAL ECONOMY SERIES

General Editor: Timothy M. Shaw, Professor of Political Science and International Development Studies, and Director of the Centre for Foreign Policy Studies, Dalhousie University, Nova Scotia, Canada

Recent titles include:

Manuel R. Agosin and Diana Tussie (*editors*)
TRADE AND GROWTH: NEW DILEMMAS IN TRADE POLICY

Mahvash Alerassool
FREEZING ASSETS: THE USA AND THE MOST EFFECTIVE
ECONOMIC SANCTION

Robert Boardman
POST-SOCIALIST WORLD ORDERS: RUSSIA, CHINA AND THE
UN SYSTEM

Richard P. C. Brown
PUBLIC DEBT AND PRIVATE WEALTH

John Calabrese
REVOLUTIONARY HORIZONS: REGIONAL FOREIGN POLICY
IN POST-KHOMEINI IRAN

Jerker Carlsson, Gunnar Köhlin and Anders Ekbom
THE POLITICAL ECONOMY OF EVALUATION

Edward A. Comor (*editor*)
THE GLOBAL POLITICAL ECONOMY OF COMMUNICATION

Steen Folke, Niels Fold and Thyge Enevoldsen
SOUTH–SOUTH TRADE AND DEVELOPMENT

Anthony Tuo-Kofi Gadzey
THE POLITICAL ECONOMY OF POWER

Betty J. Harris
THE POLITICAL ECONOMY OF THE SOUTHERN AFRICAN PERIPHERY

Jacques Hersh
THE USA AND THE RISE OF EAST ASIA SINCE 1945

Bahgat Korany, Paul Noble and Rex Brynen (*editors*)
THE MANY FACES OF NATIONAL SECURITY IN THE ARAB WORLD

Howard P. Lehman
INDEBTED DEVELOPMENT

Matthew Martin
THE CRUMBLING FAÇADE OF AFRICAN DEBT NEGOTIATIONS

Paul Mosley (*editor*)
DEVELOPMENT FINANCE AND POLICY REFORM

Tony Porter
STATES, MARKETS AND REGIMES IN GLOBAL FINANCE

Stephen P. Riley (*editor*)
THE POLITICS OF GLOBAL DEBT

Alfredo C. Robles, Jr
FRENCH THEORIES OF REGULATION AND CONCEPTIONS OF
  THE INTERNATIONAL DIVISION OF LABOUR

Ann Seidman and Robert B. Seidman
STATE AND LAW IN THE DEVELOPMENT PROCESS

Timothy M. Shaw and Julius Emeka Okolo (*editors*)
THE POLITICAL ECONOMY OF FOREIGN POLICY IN ECOWAS

Frederick Stapenhurst
POLITICAL RISK ANALYSIS AROUND THE NORTH ATLANTIC

Deborah Stienstra
WOMEN'S MOVEMENTS AND INTERNATIONAL ORGANIZATIONS

Larry A. Swatuk and Timothy M. Shaw (*editors*)
THE SOUTH AT THE END OF THE TWENTIETH CENTURY

Arno Tausch (with Fred Prager)
TOWARDS A SOCIO-LIBERAL THEORY OF WORLD DEVELOPMENT

Nancy Thede and Pierre Beaudet (*editors*)
A POST-APARTHEID SOUTHERN AFRICA?

Peter Utting
ECONOMIC REFORM AND THIRD-WORLD SOCIALISM

Sandra Whitworth
FEMINISM AND INTERNATIONAL RELATIONS

Development Administration

From Underdevelopment to Sustainable Development

O. P. Dwivedi
Professor of Public and Environmental Administration
Department of Political Studies
University of Guelph

St. Martin's Press

First published in Great Britain 1994 by
THE MACMILLAN PRESS LTD
Houndmills, Basingstoke, Hampshire RG21 2XS
and London
Companies and representatives
throughout the world

A catalogue record for this book is available
from the British Library.

ISBN 0–333–56618–1

Printed in Great Britain by
Antony Rowe Ltd
Chippenham, Wiltshire

First published in the United States of America 1994 by
Scholarly and Reference Division,
ST. MARTIN'S PRESS, INC.,
175 Fifth Avenue,
New York, N.Y. 10010

ISBN 0–312–12111–3

Library of Congress Cataloging-in-Publication Data
Dwivedi, O. P.
Development administration : from underdevelopment to sustainable
development / O. P. Dwivedi.
p.   cm. — (International political economy series)
Includes index.
ISBN 0–312–12111–3
1. Developing countries—Economic policy.   2. Sustainable
development—Developing countries.   I. Title.   II. Series.
HC59.7.D92   1994
338.9'009172'4—dc20   93–48284
                                                           CIP

To Archana and Prateek

Contents

List of Acronyms

ASPA	American Society for Public Administration
BCCI	Bank of Credit and Commerce International
CAG	Comparative Administration Group
CFCs	Chlorofluorocarbons
EC	European (Economic) Community
EE	Eastern Europe
ECLAC	(UN) Economic Commission for Latin America (and the Caribbean)
EIA	Environmental Impact Assessment
ENGOs	Environmental Non-Governmental Organizations
ESSD	Environmentally Sound and Sustainable Development
FSU	Former Soviet Union
GATT	General Agreement on Tariffs and Trade
GDP	Gross Domestic Product
GNP	Gross National Product
IBRD	International Bank for Reconstruction and Development (World Bank)
IFIs	International Financial Institutions (IBRD and IMF)
ILO	International Labour Office/Organization
IMF	International Monetary Fund
IPC	Integrated Pollution Control
LDCs	Less Developed Countries
MD	Doctor of Medicine
MNCs	Multinational Corporations
NGOs	Non-Governmental Organizations
NICs	Newly Industrialising Countries
NIEO	New International Economic Order
NPDP	National Physical Development Plan
N-S	North-South
ODA	Overseas Development Agency (UK)
OECD	Organisation for Economic Cooperation and Development
QA/QC	Quality Assurance/Quality Control
R&D	Research & Development
SAPs	Structural Adjustment Programmes
SD	Sustainable Development

SSRC	Social Science Research Council (US)
TVA	Tennessee Valley Authority (US)
UK	United Kingdom
UNEP	United Nations Environment Programme
UNESCO	UN Educational, Scientific and Cultural Organisation
USA	United States of America
USAID	US Agency for International Development
WCED	World Commission on Environment and Development

Preface

The dilemma of humanity is symbolized dramatically in the disparity among its people who are divided into two different worlds, the First World and the Third World.[1] In the First World, its citizens enjoy the tremendous gifts given by science and technology for agricultural and industrial production, for health care, communications, leisure, air-conditioned and centrally heated housing, fast and efficient transportation, public participation in the governing process, and competent and professional public service. To those in the Third World such benefits appear to be truly astounding. By contrast, the Third World appears a swarming mass of humanity living from day to day in crisis. This disparity is not only evident in income and gross national product but also in such vital fields as governance, developmental planning and administration.

The story of administration for development in the Third World is a story of various policy failures and administrative mishaps. The story starts with the independence of a few nations in Asia in 1947, and efforts to transplant the Marshall Plan experience to the developing nations. It goes on with the advent of the United Nations (UN) Development Decade in the 1960s, the continuation of this concept through the Second, Third and Fourth development decades, and efforts such as the New International Economic Order (NIEO). Disenchantment and disappointment increased in the North at the failure of its remedy: people kept on dying in thousands, squalor and wretchedness reigned supreme, the concept of democracy (as practised the North) was always upside down, and despotic rule and corrupt regimes emerged in many parts of the South. It was the global environmental crisis besetting our planet which finally brought the North back to the negotiating table with the South, in the name of protecting the planet. It is a story of failed developmental goals, told through the looking glass of administration. But more, it is about the role of the state in directing, managing and controlling the means used in and by Third World nations to achieve developmental goals; and finally, it is about the process of development administration by which those goals are supposed to be met.

The history of developmental efforts in the former colonies is less than fifty years old. With the end of the Second World War in 1945, the process of decolonisation of European empires in Asia and Africa started, beginning with the independence of India and Pakistan in 1947. The decade of the 1940s was also the period when the UN was born, and a new international monetary system was created, including the International Bank for Reconstruction and Development (IBRD). This was a period of optimism and great hope, as evidenced by the remarkable economic recovery of war-torn Western Europe by the early 1950s. Consequently, it was thought that the same experiment could be repeated in former colonies, where the people were very poor, with low per-capita income, low life expectancy, poor health and poor educational standards. The industrialised West was to provide ideas, aid, technology and technical assistance so that the ex-colonial nations could raise the living standards of their people. The key was to be economic growth, and this was to be achieved through planning. Further, it was assumed that as the West was rich, all other countries wished to be rich too; and as the West was industrialised, poor countries (which were largely rural-agricultural and natural-resource oriented) also wanted to get industrialised. It was also believed in the West that only through foreign aid (later to be called international development aid) would 'take-off' in a Rostowian manner be achieved. Further, there would be a 'trickle down' effect to the poor. Finally, the administrators of the ex-colonial countries were supposed to be change agents, nation-builders, and role-models to their people. Instead, a different administrative culture emerged.

The administrative culture of Third World countries is a product of their colonial heritage and the post-independence influence of the process of modernisation, plus a smattering of indigenous (traditional) ways of doing things. The resultant administrative style is a mixture which appears to combine the worst of these three influences. At the time of independence, the public expected that the new rulers (both political leaders and administrators) would continue maintaining law and order, justice, and fair play, augmented by social welfare and the uplifting of standards of life. The following conversation, written by George Orwell in his *Burmese Days,* between an English businessman, Flory, and an Indian medical practitioner, Dr Veraswami, illustrates that expectation:

'My dear doctor,' said Flory, 'can you make out that we are in this country for any purpose except to steal? The official holds the Burman down while the business man goes through his pockets. Do you suppose my firm, for instance, could get its timber contracts if the country weren't in the hands of the British?...The British Empire is simply a device for giving trade monopolies to the English...'

'My friend, it is pathetic to me to hear you talk so. It iss truly pathetic. You say you are here to trade? Of course you are. Could the Burmese trade for themselves? Can they make machinery, ships, railways, roads?... While your business men develop the resources of our country, your officials are civilising us, elevating us to their level, from pure public spirit. It is a magnificent record of self-sacrifice.'

'Bosh, my doctor... We have never taught a single useful manual trade to the Indians. We daren't; frightened of the competition in industry. We've even crushed various industries. Where are the Indian muslims now? Back in the 'forties [nineteenth century] or thereabouts they were building sea-going ships in India, and manning them well. Now you couldn't build a seaworthy fishing boat there.... Now, after we've been in India a hundred and fifty years, you can't make so much as a brass cartridge case in the whole continent...'

'My friend, my friend, you are forgetting.... At least you have brought to us law and order. The unswerving British Justice and the Pax Britannica.'

'Pox Britannica, doctor, Pox Britannica is its proper name.'[2]

But that 'unswerving justice' and related attributes somehow weakened over the years. Instead a new equation emerged between the political leaders and the administrators with respect to governing and administering. This aspect is examined in detail in Chapter 3.

The four decades of development efforts failed to materialise; although this matter will be discussed in Chapters 1 and 2, a brief summary of these efforts and the relationship between the North and the South is desirable here. A policy shift, in the form of a New International Economic Order (NIEO), was floated immediately after the oil crisis in 1973: the North was to provide debt relief and better access for manufacturing goods of the South in northern markets; there would be reform of International Monetary Fund (IMF) imposed conditionality and an increasing Third World voice in the international monetary system; the behaviour of multinational corporations (MNCs) would be controlled; and aid would be increased to the level of 0.7 per cent of the GNP of the North.[3] The reason why these goals remained largely unfulfilled is partly because, in developing nations, market imperfections are far more widespread than in the industrialised nations, and consequently, the control mechanisms are more complex; and partly because the market problems are rooted in broader imperfections in the governing mechanism and the weaknesses of political and administrative systems.

Then came a rude awakening in the 1980s. The 'trickle down' which was assumed to happen automatically did not occur at all, and central planning became unfashionable. While the North was enjoying an ever-increasing improvement in the quality of life, poverty remained the most serious challenge in the Third World; and as a consequence, scepticism about the effectiveness of foreign aid became widespread. Trade barriers in rich countries remained high; and the debt crisis overshadowed any feeble economic growth in the underdeveloped countries. Then the world-wide recession of the late 1980s added chaos to these problems. The final blow to the central planning system came with the disintegration of the Soviet system and its allies in Eastern Europe. By the early 1990s it became clear to the Third World that in the past decades aid proportions had fallen much below the promised 0.7 per cent of GNP; technologies which had been exported to the Third World were neither clean, nor were they meant to make Third World countries self-reliant. The commodity-dependent South has been making negligible progress, or in many cases suffering sharp reversals, in the 1980s; the net transfer of resources has been negative; and starvation, destitution, inequality, and oppression have increased.

By the early 1990s, western donors and international aid organisations started to attach a set of political and economic conditions to the granting of development aid. Political conditionality – that is, observance of human rights as practised in the West – has became one of the two primary prerequisites for development aid, which often conflicted with the right of Third World countries to chart their independent course of political development. The other condition related to capitalistic market economics or a market-friendly economic system, modelled on the American free-market philosophy. These dual conditionalities added a further burden to the political and administrative domains of the Third World.

The above is rather a bleak picture of the history of developmental efforts made in and by the Third World. However, a new ray of hope has emerged through the work of the World Commission on Environment and Development, and the Earth Summit in Brazil in June 1992, where the countries of both North and South came to realize that without mutual co-operation, this planet, and their own survival, is in jeopardy. That is why the new emphasis is on appropriate technology transfer and *environmentally sound and sustainable development* (ESSD). These two dimensions will be discussed in detail in Chapters 4 and 5.

II

The idea for such a book has been in the offing for some time, especially since a series of four articles was written by the author in collaboration with his colleague Professor Jorge Nef. The author enjoyed many hours of stimulating intellectual arguments which remained a source of much thinking, as reflected in Chapters 1 and 4 of this book. The author has in fact made liberal use of these earlier writings, for which he is most grateful to this colleague. However, the main impetus to prepare this volume came in late 1989, when the Macmillan–St. Martin's Press International Political Economy Series general editor, Professor Timothy M. Shaw of Dalhousie University, Halifax, Canada, invited the author to write on development administration.

The manuscript has greatly benefited from the helpful comments and significant suggestions made by two colleagues: Professor Vince Seymour Wilson of Carleton University, who read an earlier draft of the manuscript and made some incisive comments, and Professor Timothy Shaw, who stood by this book for a considerable time. The author is immensely grateful to have such support. He also acknowledges the following colleagues for fellowship of thought: Keith M. Henderson (Buffalo State College, US), R.B. Jain (Delhi University, India) and Dhirendra K. Vajpeyi (University of Northern Iowa, US).

Guelph, Canada O.P. DWIDEDI

Notes

1. The term 'First World' was first coined by the French demographer, Alfred Sauvy, who used *tiers monde* in the sense of the term *tiers état* meaning the 'third estate' – as existed in pre-Revolutionary France. The first estate consisted of the nobility, the second estate clergy, and the third had bourgeoisie, workers, artisans, farmers, and others. In the 1960s, this classification came to be used to denote three distinctive clusters of nations. The 'First World' consists of the industrialized nations who are members of the Organization for Economic Cooperation and Development (OECD). The 'Second World'

refers collectively to the countries of the former Soviet Union and Eastern Europe. The 'Third World' then includes the remaining nations of Latin America, Africa, Asia, and various islands around these three continents, sharing a colonial past and mostly poor, although there are some countries in this group (such as those in the Middle East and in East Asia) whose per capita income is closer to the OECD group. Further, there is another classification: 'North' and 'South' have been used to denote the 'First World' and the 'Third World' respectively. Finally, the author has used the following terms interchangeably: developing nations, underdeveloped countries, South, and the Third World.

2. George Orwell, *Burmese Days* (London: Secker & Warburg, 1961) pp. 40–1.

3. Frances Stewart, 'The Role of the South in a Chaotic World', *Development* (Journal of the Society for International Development) No. 2, 1991, p. 41.

1 The First Four Decades of Development Theory and Administration

THE CONCEPT OF DEVELOPMENT ADMINISTRATION

The term 'development administration' is of relatively recent origin.[1] The concept has been almost exclusively used with reference to the developing nations of Asia, the Middle East, Africa and Latin America. Perhaps it was first used by Donald C. Stone, although the term was popularised by Riggs and Weidner in the 1960s. But its conceptual genre has been distinctively Western. Two interconnected Euro–American traditions converge in it. One of these streams of administrative thought is the result of an evolving trend of scientific management started at the turn of the century with the administrative reform movement. The second current is the somewhat newer trend towards national planning and government interventionism which emerged as a direct consequence of the Great Depression, the Second World War and the post-war reconstruction. Events between the collapse of the international economic order in the 1930s and attempts to establish a newer one at Bretton Woods and San Francisco in 1944 and 1945 welded these two currents of administrative thought into a new synthesis which could be termed crisis management and reconstruction administration.

There are at least three historical perspectives which influenced the evolution and concept of development administration: the impact of the Great Depression and the New Deal philosophy, post-war challenges, and lessons from the Marshall Plan.

From the Great Depression and Post-war Reconstruction

With the Great Depression many of the political and economic assumptions of theories of scientific management came into question. In fact, what emerged was a radical reformulation of the old presuppositions about the role of the state in the form of Keynesianism. By the mid-1930s state interventionism had become an accepted fact in the

1

industrialised world. Massive government interventions through schemes such as the Tennessee Valley Authority (TVA) in the United States, development corporations in Europe and Latin America, and a variety of sectoral and regional programmes of economic recovery provided the necessary stimulus to refloat the troubled economies of the industrial West. In a sense, one could argue that both the New Deal and European corporatism performed a similar function: they provided a mechanism to revive the market economy. The greatest stimulus for economic recovery, however, was the way production was accelerated by using technological innovations.

Further, the war efforts during the period 1939–45 established a great number of administrative, political and economic adjustments. These were mostly related to two developments: a drastic reformulation of economic and political thinking related to the role of the state; and the introduction of a 'new' theory which gave the government a decisive managerial role in preventing the undesirable effects of the economic cycle. In addition, as an anti-cyclical function, the state assumed a leading role in procuring economic prosperity and full employment, mainly through furthering industrialisation by stimulation or direct investment. The educational, scientific and research role of the state was equally expanded, as well as the provision of social security and welfare functions.

The combination of this multitude of new tasks produced one central organisational trait: big government. In fact, the new vision of management also had to seek continuous mobilisation and participation of the public in such government projects. From a broad perspective, the role of the state was seen as correcting and rebuilding economic processes. Besides untangling bottlenecks and providing leadership in those areas where the private sector had proven ineffective, the public and the private sectors were supposed to join forces in mixed economic undertakings to increase employment and productivity. Thus, when independence came to the colonised nations, it was not surprising to see the same prescription given to them by the new coterie of 'experts' and international aid people.

Post-war Challenges

With the end of the Second World War, a myriad of new contextual factors would pose further challenges to the administrative state.

A first priority was *European reconstruction*. The American response to the challenge was the Marshall Plan. This gigantic programme was aimed at providing a massive infusion of foreign aid, thus establishing conditions for rebuilding the devastated European economies. Furthermore, it was intended to give the stimulus for accelerated and sustained economic growth to enable them both to catch up and then to achieve self-sustained growth. In the Marshall Plan, reconstruction and development were seen as two sides of the same coin and were conceptualised almost interchangeably.[2]

A second contextual factor emerging at the end of the war was a radical *transformation of the international system*. The age of imperialism came to an end and a rapid process of decolonisation began. The world political structure would shift away from a European-centred system of multi-polar balance to a rigid bi-polar one. The emergence of two superpowers with diametrically opposing economies and ideologies, coexisting in an uneasy climate of entangling mechanisms of collective defence, would characterise the new era, one of cold war.

In the realm of international organisations, the *creation of the United Nations* would have a fundamental impact in changing the fabric of international co-operation. True, the collective security role of the UN proved to be less effective than its framers dreamed. Yet a number of functional areas of international co-operation and development gave the organisation a new direction: the promotion of change through multilateral technical aid and finance. All through the 1950s and early 1960s the developmental role of the UN would become a dominant feature of the organisation and its related programmes and agencies.

Finally, and most importantly, a *'Third World' of new nations would come into being*.[3] With the exception of Latin America, the Third World was a legacy from the pre-war colonial order dominated by a handful of European powers. As the process of decolonisation started, the cold war amongst other northern superpowers moved southward. Efforts by the leadership in the new nations to transform formal diplomatic sovereignty into real economic sovereignty would become increasingly conflictual with the West, whose prosperity then still depended upon the captive markets of Africa, Asia, the Middle East and Latin America. A new breed of essentially anti-colonial and anti-*laissez-faire* nationalism would emerge in the former colonial territories, reinforcing Keynesian inclinations. And development would become the dominant issue in the Third World.

Lessons from the Marshall Plan

Since the inception of the Marshall Plan, one cold war notion had become central in Western foreign policy: prosperity was seen as an antidote to the spread of communism and other radical solutions. The Colombo Plan and President Truman's point four programme constituted the earliest Western attempts at induced development through foreign aid. On the whole, the Marshall Plan model dominated development strategies, although the complexity of the task was clearly far beyond the possibilities of Marshall-type aid. It was also obvious from the onset that underdeveloped countries, whose economies and sociopolitical structures had evolved at the periphery of a colonial system (the major difference being cultural and indigenous technological capabilities) were not in a similar position to that of war-ravished Europe, which had never experienced protracted underdevelopment. The latter required a reconstruction effort through the timely infusion of capital and technology to continue its pre-war course. The former, by contrast, were in a completely different cultural, social, political and economic situation. In concrete terms, development aid from the West was precisely directed to maintain and modernise existing economic structures, not to redirect colonial misdevelopment.[4]

Yet despite this fundamental difference, the Marshall Plan became the standard model for development. The technical aid schemes and the UN first and second development decades were nothing but expressions of this accepted doctrine. Thus, development administration emerged closely tied to foreign aid and Western formulae for development planning (based on the misleading experience of Western Europe), which were supposed to have equal (and universal) applicability in the Third World.

THE ADMINISTRATION OF UNDERDEVELOPMENT: THEORY-BUILDING

Development as a concept became an intellectual fixation in American social science in the early 1950s. Following Walter Rostow's *The Stages of Economic Growth: A Non-Communist Manifesto*,[5] the political development literature tried hard to search for the non-economic institutional conditions for accelerated, though orderly, economic growth. As far as the role of public administration in this process is concerned, two inter-related visions prevailed. One originated within the Committee on Comparative Politics of the Social Science Research Council, especially within its Political Development Group. The other vision of public administration in development came from the Comparative Admin-istration Group (CAG) of the American Society for Public Administra-tion (ASPA). Whilst both shared many assumptions, a difference of focus existed.

For those in political development, public administration was per-ceived as an institution contributing mainly to stability and *systems maintenance*. In their view, bureaucratisation was a functional condition for stability and the maintenance of legitimacy in the political order (that is, political development). As for those in comparative public administration, modern administration (that is, bureaucratic administra-tion) was essentially a mechanism for the *attainment of developmental goals*. This way, the key role of bureaucracy was seen as being a pro-cessor to provide planning and an institutional infrastructure to convert inputs of objectives, capital and know-how into developmental outputs. In the words of Donald Stone,

> Development Administration is the blending of all the elements and resources (human and physical)...into a concerted effort to achieve agreed-upon goals. It is the continuous cycle of formulating, evaluat-ing and implementing interrelated plans, policies, programs, projects, activities and other measures to reach established development objectives in a scheduled time sequence.[6]

Such characterisation of development administration emphasised the formal and technical aspects of the government machinery. Develop-mental goals were assumed to be agreed upon by the local as well as Western élites. These goals were usually referred to as 'nation-building and socio-economic development'.[7] Swerdlow has identified two inter-related tasks in development administration: institution building and planning.[8] Other authors have outlined a number of other development-oriented activities, such as the management of change, establishing an interface between the 'inner' environment and the larger intra- and extra-societal context, and the mobilisation of physical and human energies and information and their subsequent conversion into policies and actions.

Development administration is also seen as concerned with the *will to develop* the mobilisation of existing and new resources, and the cultivation of appropriate capabilities to achieve developmental goals. Thus development administration becomes an essentially action-oriented, goal-oriented, administrative system geared to realising definite programmatic values. J.N. Khosla has remarked that

> Development administration not only envisages achievement of goals in a particular area of development by making a system more efficient, it must also reinforce the system, imparting an element of stability as well as resilience to meet the requirements of future developmental challenges.[9]

Initially, development administration emphasised the developmental role of the public sector. It rested upon a series of images of development which had become predominant in western thought throughout the 1960s. Although these images did not constitute a unified development theory, a relatively identifiable paradigm did come to exist.

Five major themes can be identified. One is that development could only be attained by modernisation[10] (that is, Westernisation); in other words, by the diffusion of western values and technology. The second is that the predominant feature of development is economic development, the latter defined in terms of growth[11] (that is, the expansion of GNP per capita over a period of time). The third is that quantitative change (economic change) would produce a critical mass leading to qualitative changes. Sequentially, economic growth would bring about social changes which, in turn, would bring about political development. Structurally, an expansion of wealth in the hands of an investor élite would trickle down, bringing generalised prosperity and a higher standard of living. The fourth theme is that the process of development historically entails the movement of societies between a traditional agrarian stage of underdevelopment and that of development after the take-off stage (industrial). All societies are postulated to be developing or in transition between these two poles. Furthermore, all nations are said to have been at one time underdeveloped. For instance, today's industrial societies were once agrarian and feudal. By sheer cumulation or induced changes in their social structures and value systems they became developed. What is more, the paradigm postulates that once a region becomes developed, capital, technology and ideas would, in turn, bring development to other areas. The fifth main developmental underpinning of development administration is the emphasis on harmony: 'stable and orderly change'.[12] Development in this context is perceived not only as the attainment of change but also, and mainly, as adaptation and systems-maintenance.

Administrative Modernisation and Institutional Transfers

One of the lessons that the industrial nations learned from the depression, the war and the Marshall Plan was that reconstruction and recovery could be dramatically accelerated by improved management and organisation. The prevailing mood of administrative analysts and practitioners was that the vast amount of experience in reconstruction management and organisation could be adapted to suit the specific developmental needs of the post-colonial world. In fact, development administration was seen to be a mutation of colonial administration, to be obtained by injecting development goals and structures into the old core of civil servants. The task of the developed countries was perceived as the creation of external inducements to change[13] through technical assistance and transfers of technology and institutions. Such a strategy of Westernisation was directed to both the administrative machinery and to the whole national community. The most fundamental ingredient in the process under induced development would be inputs of foreign know-how and capital (in the form of either aid or investment). A number of techniques such as programme planning, community development and personnel management, popularised during this era, reflect this bent for external inducement towards modernisation and Westernisation.[14]

Needless to say, such an approach requires technological, economic and institutional diffusion from the developed towards the depressed, devastated or underdeveloped regions. Moreover, diffusion of modern know-how is perceived as being value-free and culturally neutral. A related assumption is that institutional imitation is bound to produce results similar to those obtained in the developed world: efficiency, increased rationality and the like. At a very general level, the diffusionist trait in development administration highlights the fundamental link between administrative efficiency and the consolidation of bureaucratic characteristics referred to earlier. It was assumed that the more developed (that is, bureaucratic and Western-like) an administrative system became, the greater the likelihood that it would have developmental effects.

THE FIRST THREE DECADES OF DEVELOPMENT: 1950 TO 1970

Although from the UN viewpoint the first development decade began from 1960, in reality it got started in the 1950s with President Truman's point four program and the Colombo Plan. This was a decade of optimism, expectations and the establishment of international aid agencies in various industrialized countries. The decade of the 1960s was an era of general prosperity and also one of pervasive intellectual optimism throughout the world. The inspiration of visionaries and the belief that modernisation and technology would surmount any obstacle to human progress were the orders of the day. It was thought that, with sufficient foreign aid and a revamped administrative system, Third World countries would follow, if not overtake, the industrial and technical levels of the West.[15] There was confidence that an administrative state would triumph with the help of new tools of development administration. Examples of reconstruction and rapid recovery in Western Europe and Japan were used to strengthen this belief. And so, when multilateral foreign aid programmes were inaugurated in the 1950s, western social scientists, administrators and social engineers envisioned a world-wide utopia: new societies, new frontiers, national integration and global development through technical co-operation. Several so-called wars were fought with the emerging administrative hardware: the war on poverty, the war on underdevelopment and the cold war. Administrative and military modernisation – both closely related developments – became the operational mechanisms for the preservation of postcolonial western ascendency over the developing areas.[16] But the expected administrative paradise did not materialize.

The early 1970s marked a rude awakening to the inadequacies of the developmentalist paradigm of public administration to cope with urgent problems. The crisis of development administration in this decade became one of identity and purpose, with seemingly devastating effects on the entire field of public administration, in the North as well as the South. Assumptions, methodology and focus became increasingly irrelevant. In fact, after its accelerated growth in the 1960s, development administration apparently plunged into the depth of an intellectual depression.

The second decade of development had lost its impetus by the late 1960s; a spirit of frustration and despair with development administration and with development in general had set in. For one thing, it appeared evident that externally induced modernisation had failed to eradicate the basic problems of underdevelopment which it purported to solve. Whilst some significant increases of GNP had indeed taken place, poverty, disease and hunger had either worsened or remained unaltered. The same could be said of the growing gap between the rich and poor nations, not to mention that between different social strata within nations. In many regions, incremental reformism had failed to create a more equitable socio-economic order and had proven to be an ineffective antidote to radical change. In fact, frustrated reformism had fuelled a revolution of rising frustrations. Large on the horizon loomed the Indo-China experience. It demonstrated that over-administering was neither an efficient nor an effective insurance against revolution.

Development administration had equally shown poor results on the home front. In the mid-1960s several crises occurred in the West: urban decay, social upheaval, protests and a deep questioning of institutions.[17] The end of ideology had not come, just the end of consensus and liberalism. The discipline had to

> grapple with the effects of the economic and political crises of the domestic and international fronts ... Instead of affluence, growth and optimism, the administrative environment [was] being increasingly characterized by scarcity, stasis (or decline) and lowered expectations.[18]

Two basic trends are related to this decline. First, at the level of praxis a great deal of the international and domestic development efforts had proven less than impressive. Secondly, at a conceptual level, the failure of development and reformism in general resulted in an expanding analytical void. Increasingly, the objectives, methodologies and even epistemologies and value assumptions of western social science, especially political, administrative and development theories, would undergo fundamental questioning.

The Crisis of Development Strategies

The failure of the majority of developmental efforts is obviously too complex to be dealt with here in any detail. It should be remembered, however, that by the late 1950s and early 1960s development was taken for granted – all that needed to be done was to establish adequate conditions to induce its arrival. This was the midwifery task of development administration. Development administration – the mobilisation of people and resources towards achieving a modern society – would, in turn, result from administrative development (in other words, the modernisation of the administrative machinery). This view dominated both development assistance abroad and domestic regional and sectorial policies in the developed world.

By the 1970s development could no longer be taken for granted, while at the same time the tendency towards decay had become noticeable in both the North and South. Events like the energy crisis, the growing economic recession in the major industrial countries, and a crisis of liberal democracy in the early 1970s dampened most traces of earlier optimism. An increased contradiction between market economies and market policies would also serve to undermine some of the once-considered-persistent civic traditions of Western pluralism. This contradiction was rooted in the severe limitations, even of continuous economic growth to reduce social antagonisms. Its implications – the fiscal crisis of the state and a manifest trend towards stagnation and political stalemate – obviously constitute a new context for public administration.

The Field in Search of a Discipline

By the late 1960s development administration had suffered the simultaneous and reinforcing impact of two epistemological traits. One was a generalised trend towards rejection of the functional–structural, systemic and behaviourial fads in social science. These vogues had indeed provided a great deal of the glue to construct comprehensive conceptualisations such as those of Riggs, Eastman and others.[19] The second was more ingrained in development administration proper, and resulted in what could be called a 'Tower of Babel' syndrome. As Caiden has vividly portrayed:

With development becoming a more magical word every moment and with more resources available for the study of anything about development, development administration became a catch-all for idiographic applied social scientists and nomothetic theorists. Development administration took off into modernization, nation building, social change, industrialization, cultural anthropology, urbanization, political ecology and anything else that seemed to promise help for policy makers in developing countries.[20]

In the extreme concern for contextualism and interdisciplinary analysis, development administration has greatly blurred the conceptual focus of public administration. What had been gained in richness has been lost in simplicity and precision. Furthermore, with an expanding interest in the context, a great deal of the very distinctively prescriptive intent of public administration has faded away. It should be remembered here that the very origins of the discipline were tied in with instrumental considerations. To affect the environment, not solely to explain it, was a fundamental component in the entire history of the discipline. The puristic, value-free stand of most of the logical–positivistic paradigms finally separated administrative theory from its professional practice. The problem was that neither the CAG of the ASPA nor the Committee on Comparative Politics of the SSRC had a unified vision of what the field was.[21] A mixture of genuine enthusiasm, naïvety, and 'bandwagonry' seem to have played a major part in these movements. In fact, clear agendas or strategies were never instrumentalised. Further, whilst the CAG emphasised development and the attainment of change through administration, the comparative politics group focused on bureaucracy as an institution contributing to the system's maintenance. Thus, it was not surprising to see that a crisis of purpose and identity set in. Consequently, by the late 1960s, development administration had become a field in search of a discipline.

The Crisis of Development Theory

Perhaps the single most important intellectual crisis in development administration was the shattering, by the late 1960s, of development theory. Deep controversies followed the apparent developmental failures of the decade, and a number of radically different conceptions of development, produced mostly by Third World scholars, came to be presented. These alternative theories were not merely reactions either to the traditional school or its practical failures; some had been in existence for a good number of years before they made their entry into western academic circles. Their academic recognition, however, did not mean a new theoretical synthesis. Quite the contrary, development theories polarised into two identifiable ideological camps: one traditional, the other radical. The overall effect of the radical critique was to unmask the purported value-free posture of the conventional wisdom. The very concept of development was severely questioned. Whilst for traditionalists it still meant economic growth, increasing numbers of analysts had begun to define it in terms of human values: quality of life, distribution, satisfaction of basic needs and so on. Development 'for what', 'to what' and 'for whom' were not merely rhetorical questions; they highlighted the fact that development entails prior normative considerations and value choices. Particularly important here are the distinctions of development as 'having' with the emphasis on material progress as opposed to 'being' (intrinsically related to human development as a whole) and the fundamentalist conception of 'development as liberation'.[22]

The vision that development spreads from the developed to the underdeveloped in a contagious manner had been refuted on both logical and empirical grounds. Diffusionist schemes at the level of international and national development, radical critics argued, were oblivious to the fact that the processes of development and underdevelopment were dialectically interrelated.[23] That is, development and underdevelopment are not two poles of a continuum but are inextricably linked as one single historical process. From the dependency perspective, modernisation (westernisation) need not be looked upon as a vehicle for development but, paradoxically, as a major contributor to underdevelopment. Modernisation, then, may not lead to economic growth. Even if it does, such growth need not lead to economic development. Nor will it necessarily result in social and political development.

Impatience with developmentalism was growing in other quarters as well. Conservative critics of the neo-liberal, developmentalist model rejected the distributive overtones of development theory. Instead, they emphasised growth and stability as the sole objective concerns of development theory. They looked in antipathy towards government planning, mixed economies and interventionism.[24] In their view, the public sector had a very limited direct role to play in development, because the function of the state was to create a stable and fiscally sound environment in which private business could engage in the unmolested pursuit of growth. Development administration and induced development schemes, therefore, constitute only artificial distortions to objective market forces. Overall, it is largely these neo-classical theories that dominated American developmental thinking during the Nixon and Ford years and after.

Summary

The failure of development administration strategies was a consequence of factors outlined above. The tension between development administration, which emphasised mobilisation, and administrative development, which was centred in the capacity for social control, lay at the core of its failure. Nevertheless, the crisis suffered by the discipline was not severe enough to make it disappear altogether.

THE FOURTH DECADE OF DEVELOPMENT: THE 1980s

By mid-way through the third development decade (1971–81) then, the very foundations of the development administration paradigm were severely in question. Not only was its usefulness in doubt in the Third World itself, but an intellectual crisis had set in among students of development administration in the West. The gap between the centre and the periphery was widening rather than narrowing in both relative and absolute terms. Instead of development and nation-building, turmoil and fragmentation proliferated throughout Africa, Asia, the Middle East and Latin America. Urban crises, energy crises, cessation of growth, unemployment, and breakdowns of public institutions and public morality – all had a dampening effect on the early optimism about the ability of the First World administrative technology to solve problems everywhere.[25]

During the fourth development decade, the New International Economic Order (NIEO) became an important new symbol in the development arena. Its demand for a basic realignment of the world economy, through changes in trade, aid and technological transfer, was appreciated but generally ignored by the richer donor nations. There was in fact no global consensus concerning NIEO objectives, and some commentators felt that it might even harm certain countries. At the same time, the NIEO represented a basic-needs strategy, attempting to redirect the terms of reference of the development debate: that is, the strategy would provide the essential necessities for people, yet worry less about such traditional indicators of growth as GNP per capita. While the World Bank and the International Labour Organisation (ILO) symbolically endorsed this approach to development, the monumental change in the world system demanded by the NIEO did not occur. Further, in the absence of a common strategy adopted by both the West and the Third World, the NIEO also went the same way as the other concepts which had emerged earlier.

Since the 1980s, another important factor which has broadened the scope and spectrum of Third World problems is the withering away of Eastern European and Soviet Union communist states, and the entry of at least some of them, especially in Central Asia, into the realm of the Third World. Yet, although the Eastern Bloc nations may be considered 'poor' for the time being, many will not remain so for long. Akin to the economic recovery of Western Europe after the Second World War, many in Eastern Europe and the former Soviet Union, except Central Asia and parts of the Balkans, may leave the rest of the Third World nations behind.

Development administration cannot be divorced from either political economy or a theory of development. The core assumption here is the identity between development and modernisation, the latter understood as westernisation.[26] The function of development and administration was chiefly that of a midwife for western development, creating stable and orderly change. The West would produce the external inducements thought necessary to promote such change.[27] Development-mongering became synonymous with reform-mongering. The latter was always oriented to preventing discontinuities in the mode of production, élite structure and international alignments. The implicit goal, it now seems, was to modernise yet not to uproot the existing structures and processes, thereby keeping the poor, mono-producer and backward societies locked into a cycle of dependence and underdevelopment. Put in rather simplistic terms, the prevailing paradigm assumed that the problem, whatever its nature, remained with and at the periphery. Yet, the solution was always locked in the centre. In an extreme way, the prevalent attitude was not only that 'they' had the problem, but that they *were* the problem. Conversely, the West was postulated both to *have* and to *be* the solution. Traditional societies had to be saved, if not from the appeals of communism, then at least from themselves.

Further, the orientation in these newly independent nations was very much pro-status quo. The quick metamorphosis of colonial European officers in Africa and Asia into their 'new' positions as UN development administrators is a case in point. This was hardly more than a relabelling of old colonial bottles of a law-and-order vintage, with the managerial sticker required to develop the former colony. A similar situation occurred with the native cadres formed during colonial days to administer on behalf of European interests. Personnel, procedures, habits, structure and values inherited from their colonial past were hastily wrapped in the garments of nationhood.

A developmental creed soon emerged which posited that in order to attain development, a country's administrative structure should conform to the standards of the most advanced industrial societies. The key issue, then, was the transformation of the existing traditional machinery into the new entity. This was to be accomplished through administrative development: the modernisation of the public service machinery through external inducement, transfer of technology, and training by foreign so-called experts. For this task, there was already a neat prescriptive model to be found in western tradition. This tradition was based on the dichotomy between politics and administration; it was a system that relied on hierarchy, unity of command, political neutrality, recruitment and promotion on the merit principle, public service accountability, objectivity and ethical probity. While external sources placed a great insistence on sustaining these values in the Third World, in reality these principles were supplemented by existing traditional methods. Thus, a parallel value system gained currency: western models were set up and existed simultaneously with the traditional economy and black market. Rarely were the principles of development administration, as recommended by the experts from the West, questioned; as a matter of fact, these were generally accepted at face value by native élites, especially where a relatively smooth transition to nationhood took place.

The post-independence political and bureaucratic leaders rapidly moved to replace the colonialists.[28] Western education, as formally imparted in these countries and as continued by returning graduates from the West, was perceived as a tool both for personal advancement within the organisation and for acceptance by similarly trained professionals in the West. Thus it is not surprising to find that an administrative machinery took shape that was incapable of implementing developmental goals, particularly in dealing with poverty and scarcity. In spite of much rhetoric, the emergent administrative systems tended to be imitative and ritualistic. Practices, styles and structures of administration generally unrelated to local traditions, needs and realities succeeded in reproducing the symbolism, but not the substance, of a British, French or US administrative system. Even where a relatively large contingent of trained functionaries existed, such as in India, Pakistan, Kenya, Nigeria, and Ghana, a continuation of colonial administrative culture prevailed. Yet at the same time a massive dose of political interference in the way of doing things stifled developmental initiative. Confronted with an ineffectual developmental bureaucracy, the western solution was to call for even more administrative development. This was also the preferred option of the local elites. Technical solutions about means were more palatable than the substantive political decisions needed to bring about real socio-economic change. Administrative reorganisations and rationalisations for the sake of abstract principles soon became the ends rather than the means of development administration.

During the 1980s it became a pastime among international aid experts to talk about the 'Chinese model' often out of context and in almost mythological terms. Even though it is somewhat paradoxical, the World Bank and the IMF – the upholders of western ways and ascendancy over the Third World[29] – became proponents of elements in the Chinese solution, such as an emphasis on basic needs. The partial incorporation of the rhetoric of the model has often resulted in an idealised, distorted and trivialised representation of an extremely different experience.[30]

The developing nations were disillusioned when they found that, instead of being recipient of capital *from* the West, they were forced to make net transfers of their meagre resources *to* the West. In order to service their debts, several developing nations became virtually bankrupt. Restrictive trade practices prohibited the poor from exporting their products. The problem was further compounded when commodity prices fell to the lowest level for fifty years, while the prices of manufactured products kept rising. It is no wonder that starvation, destitution, inequality and oppression have continued in most of the developing nations. The Golden Age of the 1950s had turned into the Age of Pessimism and Disillusion by the 1980s. Nevertheless, some rays of hope came from the world-wide awareness displayed at the Earth Summit in 1992, which demonstrated that no one group of nations could keep on 'progressing' while the majority remained hungry and poor. It may sound moralistic to argue that it is the duty of the North to help close the development gap with the South, but is there another way out? Until the 1990s, the South adopted a weak approach to negotiations with the North. This means that new approaches are required to North–South relations.

CONCLUDING OBSERVATIONS

We have for over a century been dragged by the prosperous West behind its chariot, choked by the dust, deafened by the noise, humbled by our own helplessness, and overwhelmed by the speed... If we ever ventured to ask, 'progress towards what, and progress for whom', it was considered to be peculiarly and ridiculously oriental to entertain such doubts about the absoluteness of progress.[31]

The sentiment above was articulated by Rabindra Nath Tagore, the Nobel Laureate, in about 1908; it seems that Tagore's thought is even more relevant for the 1990s than it was then. The present chapter has tried to answer one of the questions posed by Tagore: progress for whom? It seems that while the Third World nations have witnessed an increase in their poverty level, the West has become more prosperous. At whose cost?

This chapter has surveyed a difficult terrain whose topography is continually changing. The author has grouped the various approaches to development theory and administration into three broad categories: counter-insurgency and status-quo oriented Western models, revolutionary efforts made by some countries, and the traditional reforms tried in a few instances. In the light of the preceding analysis, one should now ask some fundamental questions: Is the present theory and practice of development administration, its basic concepts, assumptions and values, still relevant in the 1990s? If not, and if we believe that a serious crisis has overtaken the discipline, should we consider developing a new strategy more suited to tackling the satisfaction of people's basic needs, the eradication of poverty and the protection of human dignity?

With regard to the first question, if we mean by development administration the western approach, the answer is clearly no: the conventional wisdom has been an unqualified failure. There is a need to develop an alternative strategy based on the wealth of material available on the Third World which deserves to be studied and systematically analysed. Moreover, if we judge the question of relevance as a synthesis between Third and First World paradigms, the answer is, once again, negative. The current crisis of development administration is precisely a consequence of the inability to incorporate the substance of other non-western developmental experiences into the prevailing conceptual mould.

In prescriptive terms, a much-needed intellectual synthesis is still lacking. However, such an exercise is beyond the scope of this chapter. It is important, nevertheless, to suggest a number of possible issues around which a new conceptual approach could be built. These major issues can be grouped into three categories: accountable development administration; the role of values in development administration; and the emergence of fundamentalism as a new development ideology.

Accountable Development Administration

Conventional development administration assumed that once administrators were assigned their tasks, the system would automatically create its own chain of accountability. Developmental objectives, which were painfully sketched in long-range plans and through administrative directives, were assumed to be fulfilled once the field administrators were given the authority and necessary resources to implement them. However, no serious effort was made to exact accountability for programme objectives to determine whether the scarce resources had been used efficiently and effectively. Developmental administrators, while handling the mammoth task of nation-building (with seemingly unlimited authority to redistribute and regulate the economy), often opted for an easy way out by making sure that their actions were protected by the necessary paper work. This, naturally, resulted in an over-dependence on rule books and red tape – a phenomenon widely prevalent in all Third World countries.

This preoccupation has created a bureaucratic environment of distrust among public servants, which has inhibited the creation of a milieu for effective, responsible and accountable administration. Unless the administrators are held accountable for their decisions, and unless they are given credit for satisfactory results or reprimanded for failures, development administration cannot achieve the necessary credibility and effectiveness among the masses. The main issue in Third World societies is not merely the availability of appropriate technical and administrative skills, but rather the creation of a politico-administrative environment oriented toward securing basic necessities for the populace. This means harnessing human capital and energies (which are in abundant supply in many Third World countries) toward broad-ranging social objectives primarily related to basic needs. An accountable development administration would then strive toward serving those needs by becoming responsible and responsive to societal needs. Political accountability will, then, precede development administration accountability.

Value-Free Development Administration?

As mentioned earlier, one of the basic assumptions of orthodox development administration has been the belief in a scientific, neutral administration. The Western myth – that value laden decisions are in the domain of politicians while public servants merely implement these decisions, with no room to influence policy choices – was transplanted to Third World societies. While this may have been plausible in the public bureaucracies of the western countries with a tradition of the separation of powers, in reality, in the South, developmental administrators were never free from value-laden decisions. With the concentration of authority and discretionary power in the hands of administrators, and in the absence of appropriate social and political accountability for the use of such enormous state power, public servants found themselves playing an almighty role in the allocation of scarce resources. Thus, among the various misapplications of the Weberian concept of bureaucracy in the Third World, the value-free administrative system appears to be one of the most counter-productive myths that has been transplanted. It has created an environment in which administrators hesitate to express openly their views on policy issues on the pretext of civil service neutrality.

In fact, what is needed is a cadre of administrators who are willing to state their opinions and values on programmes and projects. Expressing

individual values does not mean that public servants ought necessarily
to become committed party functionaries. Based on the record of success of some Third World countries, it appears that the attainment
of developmental goals is strengthened where the administrative
machinery has freedom to express its values and beliefs without fear or
favour.

Influence of Culture: The Missing Link in the Paradigm

Theories of development have been based on three fundamental variables: improving technological capability, economic development, and
administrative changes. If treated at all, culture as a factor has been
taken to be an epiphenomenon: it was never seen as a central or even a
secondary factor. From this author's perspective, human beings and
their cultures are inseparable. For example, countries such as Iran in the
post-Shah era, Pakistan, Bangladesh, Egypt and Algeria, to name only a
few where religion continues to wield considerable influence, are under-
going rapid changes in many aspects of their culture. Even in a sup-
posedly secular nation such as India, the influences of religion, caste and
family ties have shown a remarkable capacity to adjust to democratic
politics. Some may treat this aspect as one of the growing influence of
fundamentalism; but it is by no means anti-developmental in its impact.
Rather, this aspect involves the juxtaposition of traditionalism and mass
arousal. It generally involves a rejection of western-style modernisation,
especially its alien values. It requires the existence of a very cohesive
and traditional nationalistic elite that is capable of controlling or
hegemonising the political and ideological discourses.

It should be noted that not all aspects of a culture promote development. It is up to the vision of leaders of these nations to take advantage
of the favourable elements while neutralising the unfavourable ones.
Any development programme should be based on a deep understanding
of the culture of the people whose lives are going to be changed 'for the
better'. The need to appreciate the cultural context in the developmental
process becomes much more crucial when developmental plans and
programmes are either imposed from abroad or are part of a so-called
structural-adjustment-dictated design. We already know that the traditional forces may have, at least partially, a developmental role to play
by raising central issues of the philosophy and ethics of development.

That is why this cultural aspect becomes a major factor in appreciating delay, decay, or progress in developmental efforts.

The challenge before international bodies and aid-giving agencies now is to make a difficult but not impossible mid-course correction in their approaches. It should be noted that as the cultural and socio-political nature of countries differs, so Third World countries cannot rely upon one single model of development. Each country will have to innovate its own strategy of development by borrowing, grafting and improvising upon its inherited indigenous capabilities. Before that, however, Third World societies must become intellectually self-reliant by charting their own theories and methodologies of development.

Summary

Whereas the previous four development decades emphasized modern-isation through the transfer of technology (both ideas and tools) assisted by foreign aid, as well as the NIEO, it was acknowledged in the early 1990s that the fifth development decade, if it is going to be relevant, must change its focus and strategy to include such key goals as sustain-able development, human resource development, empowerment of specific groups, and removal of poverty. Development administration could provide the impetus for the achievement of these core objectives effectively and forcibly. This aspect is examined in the following chapter as well as at the end of the book.

Notes

1. This section is drawn from J. Nef and O.P. Dwivedi, 'Development The-ory and Administration: A Fence Around an Empty Lot?' *Indian Journal of Public Administration*, Vol. 27, No. 1, January–March 1981, pp. 42–66.

2. Clyde Sanger, 'Pearson's Eulogy', *International Journal*, No. 325, 1969–70, p. 179.

3. Irving Louis Horowitz, *Three Worlds of Development. The Theory and Practice of International Stratification* (New York: Oxford University Press, 1966) pp. 3–14.

4. C.R. Hensman, *Rich Against Poor. The Reality of Aid* (Harmondsworth: Penguin, 1975) chapter 3, *passim*.

5. Walter W. Rostow, *The Stages of Economic Growth: A Non-Communist Manifesto* (Cambridge, Mass.: Harvard University Press, 1960).

6. Donald C. Stone, 'Tasks, Precedents and Approaches to Education for Development Administration', in Donald C. Stone (ed.), *Education for Development Administration* (Brussels: International Institute of Administrative Sciences, 1966) p. 41.

7. Milton D. Esman, 'The Politics of Development Administration', in John D. Montgomery and W.J. Siffin (eds), *Approaches to Development, Politics, Administration and Change* (New York: McGraw-Hill, 1966) pp. 69–70.

8. Irving Swerdlow, *The Public Administration of Economic Development* (New York: Praeger, 1975) pp. 15–19.

9. Quoted in J. Nef and O.P. Dwivedi, 'Development Theory and Administration: A Fence Around an Empty Lot?' *Indian Journal of Public Administration*, Vol. 27, No. 1, January–March 1981, p. 50.

10. Swerdlow, *Economic Development*, p. 345.

11. Gerald Meier (ed.), *Leading Issues in Economic Development* (New York: Oxford University Press, 1970) p. 7.

12. Susanne Bodenheimer, 'The Ideology of Developmentalism: American Political Science's Paradigm – Surrogate for Latin American Studies', *Berkeley Journal of Sociology*, Vol. 15, 1970, pp. 95–137.

13. Ralph Braibanti, 'Transnational Inducement of Administration Reform: A Survey of Scope and Critique of Issues', in Montgomery and Siffin (eds), *Approaches*, pp. 133–83.

14. Bernard Schaffer, *The Administrative Factor* (London: Frank Cass, 1973) pp. 244–5.

15. Garth N. Jones, 'Frontiersmen in Search for the "Lost Horizon"', *Public Administration*, Vol. 36, No. 1, January–February 1976, p. 99.

16. Brian Loveman, 'The Comparative Administration Group, Development and Anti-Development', *Public Administration Review*, Vol. 36, No. 6, November–December 1976, pp. 6–20.

17. H. George Frederickson, 'Public Administration in the 1970s: Development and Directions', *Public Administration Review*, Vol. 26, No. 5, September–October 1976, pp. 564–5.

18. Carl W. Stenberg, 'Contemporary Public Administration: Challenge and Change', *Public Administration Review*, Vol. 36, No. 5, 1977, p. 507.

19. See, for example, Fred Riggs, *Administration in Developing Countries* (Boston: Houghton-Mifflin, 1964); Milton Eastman, 'The Politics of Development Administration', in Montgomery and Siffin (eds), *Approaches*, pp. 59–112.

20. Gerald E. Caiden, *The Dynamics of Public Administration: Guidelines to Current Transformations in Theory and Practice* (New York: Holt, Rinehart and Winston, 1971) p. 267.

21. Fred Riggs, 'The Group and the Movement: Notes on Comparative and Development Administration', *Public Administration Review*, Vol. 36, No. 6, November–December 1976, pp. 648–50.

22. Denis Goulet, *The Cruel Choice, A New Concept in the Theory of Development* (New York: Athenaeum, 1973), pp. xii–xxi.

23. See, for example, Susan George, *How the Other Half Dies: The Real Reason for World Hunger* (New York: Penguin Publishing Company, 1976) p. 94.

24. See, for example, Samuel Huntington, *Political Order in Changing Societies* (New Haven: Yale University Press, 1968) pp. 4–6; Milton Friedman, *Capitalism and Freedom* (Chicago: University of Chicago Press, 1962) pp. 7–36.

25. This section draws from O.P. Dwivedi and J. Nef, 'Crises and Continuities in Development Theory and Administration: First and Third World Perspectives', *Public Administration and Development*, Vol. 2, 1982, pp. 59–68.

26. Swerdlow, *Economic Development*, p. 345.

27. See, for example, Braibanti, 'Transnational Inducement,' and Schaffer, *Administrative Factor*, pp. 244–5.

28. See L. Kooperman and S. Roseberg, 'The British Administrative Legacy in Kenya and Ghana', *International Review of Administrative Sciences*, Vol. 43, No. 3, 1977, pp. 267–72.

29. See N. Girvan, R. Bernal and W. Hughes, 'The IMF and the Third World: The Case of Jamaica', *Development Dialogue*, Vol. 2, 1980, pp. 113–55.

30. See M. ul Haq, 'An International Perspective on Basic Needs', *Finance and Development*, Vol. 17, No. 3, 1980, pp. 11–14.

31. Quoted in George Axinn's 'Sustainable Development Reconsidered', *Development* (Journal of the Society for International Development), No. 1, 1991, p. 120.

2 Development Administration in the Fifth Decade

This chapter will survey the impact of more than four decades of involvement by indigenous public administrators, often with external aid, in the process of nation-building and socio-economic development. It also seeks to assess the situation of administrative culture and morality which exists in such nations. And finally, it will project dilemmas and responses for the 1990s, looking towards the twenty-first century.

THE FOUR DECADES OF DISAPPOINTMENTS

When the age of imperialism came to an end after the Second World War and the rapid process of decolonisation began, it was thought that, with sufficient foreign aid and a revamped administrative system, these ex-colonies would closely follow, if not altogether achieve, the industrial progress of the West. Aid and administration for development became mechanisms to fight the war on underdevelopment.[1] With the exception of Latin America, the Third World was a legacy from the pre-war colonial order dominated by the European powers. As the process of decolonization began, the efforts of the leadership in the new nations transformed formal diplomatic sovereignty into conflict with the West. A new breed of essentially anti-colonial and *laissez-faire* nationalists emerged in what were formerly colonial territories.[2]

In China, India, the Middle East, Southeast Asia, and West and East Africa, a new wave of expectant peoples strained the emerging neo-classical order while tearing down the remnants of the imperial system. Development had become the dominant issue in the Third World. In the 1960s the West responded to this development challenge in a number of ways. The first was to conceptualise the notion of development administration by blending all necessary elements of human endeavour with financial and material resources in order to achieve developmental goals that were generally recommended by the western experts. Development administration was seen as concerned with the will to mobilise existing and new resources, and to cultivate appropriate capabilities to achieve developmental goals. Thus development administration became an essentially action-oriented, goal-oriented, administrative system geared to realise definite programmatic values. The task of the North was perceived to be the supplier of external inducements to change through technical assistance and transfers of technology and institutions.[3]

Development administration was supposed to be based on a professionally oriented, technically competent, politically and ideologically neutral bureaucratic machinery. It was to retool foreign aid (both in the form of funds received and ideas suggested) and also to act as a main instrument for nation-building by transforming the inputs received into development outputs. The ostensible output was modernisation, preceded by institution-building and adaptation of the indigenous bureaucratic machinery to undertake developmental tasks. Thus, the developmental bureaucracy, as a spearhead of modernisation, was seen more as an adapter than as an innovator.

But later events betrayed such expectations. Administrative reforms tended to have the long-run consequence of strengthening the old framework. Moreover, given the acute shortage of qualified human resources to manage the public service, innovations often originated with expatriates and foreign experts, or were simply cosmetic structural changes that brought no real alteration in the status quo. Confronted with an ineffectual developmental bureaucracy, the western solution was more administrative development. Technical solutions about means were more palatable than much-needed substantive political decisions to bring about real socio-economic change.

By the end of the 1970s it was clear that something had gone wrong. The western style of economic progress was obviously not forthcoming; instead, the quality of life in many Third World countries was declining. In place of orderly change, turmoil and fragmentation proliferated throughout Asia, the Middle East, Africa and Latin America. Moreover, the recession of the later 1970s and early 1980s curbed the enthusiasm of developmental strategists of the West and created doubts in the Third World about the invincibility of western wisdom. Thus, the failures in development we see in most of the Third World nations have resulted partly because of their own inadequacies, but also because

(1) for many years western scholars have been unable to include the non-western contributions to developmental studies;

(2) ethnocentricism and ignorance in the West have continued to overshadow the need to appreciate the role of local tradition, culture, religion, and style of governance;

(3) the infusion of foreign aid has failed to enhance the quality of life or satisfy basic needs.

Obviously, this loss of confidence has also affected the discipline of development administration. This confusion may continue because variables, such as the impact of the colonial legacy on the administrative style, emerging administrative culture and morality, and challenges facing Third World rulers, are yet to be subsumed fully in the paradigm and assumptions defining the field.

Administrative Heritage

After achieving independence, the efforts of the Third World leadership were concerned to transform formal diplomatic sovereignty into political independence and economic viability. A wave of students left the developing nations for higher education in the West, to return with vigour and enthusiasm for the things that were western. Several developing countries started experimenting with notions of development administration which were to blend human endeavour with financial and material resources in order to achieve developmental goals. Such characterisation of development administration emphasised the formal and technical aspects of government machinery.

Two interrelated tasks in development administration were identified: institution-building and planning. Development administration was also seen as being concerned with the need to develop and mobilise existing and new resources, and to cultivate appropriate capabilities to achieve developmental goals. Thus development administration became an essentially action-oriented administrative system geared to realizing definite programmatic values. The developing countries did not realize, however, that the most fundamental ingredient in the process of induced development was going to be the input of foreign expertise and capital. A number of techniques popularised during this era, such as five-year planning, community development and administrative reforms, reflected the developing countries' preference for external help for modernisation and westernisation. A related perception was that institutional imitation was bound to produce similar results to those obtained in the West. It was also felt that the more developed and western an administrative system became, the greater the likelihood that it would produce developmental effects.

What was missing from the expected picture-perfect imitation in the Third World was the necessary set of conditions for bringing about a number of social, economic, cultural and political changes. These include an expanding economic base, professionally trained human resources, political maturity, cultural secularisation, a relatively open society and a strong political superstructure capable of governing. Administrative reforms tended to strengthen the old framework, bringing no real alteration in the status quo.

THE FIFTH DECADE: THE 1990s – DILEMMAS AND RESPONSES

From anyone's standpoint, the decade of the 1990s up till 1993 does not represent an improvement over the previous forty years. Administrative culture and morality remain problematic, and improvements within the administrative infrastructure are few and far between. Several long-established assaults on administrative inadequacy were still to be found at the beginning of the 1990s, but also one newer approach. The former were in the nature of orchestrated campaigns supported by top political leadership; restructuring of administrative organizations including the creation of new agencies; the use of control mechanisms, increased penalties, anti-corruption boards, and the like; various training programmes; and streamlining of procedures. In its early years, comparative public administration and its protégé, development administration, were closely associated with these activities. The newer approach is to cut back the scope of government activities through privatisation (particularly of para-statals), deregulation, decentralisation, and similar efforts, all of which may be subsumed under the rubric 'de-bureaucratisation' – in other words, a reduction in the scope, or at least in the rate of growth, and the streamlining of procedures of the centralised administrative apparatus in the public sector. This is the dominant trend of the 1990s.

Although this chapter is concerned with the administrative side of government operations, recognition must be given to the context in which administration occurs and statecraft is practised. This context is often so mixed up with administration that there appears to be hardly any meaningful distinction between policy formulation and policy implementation. Bureaucrats at all levels, from the ministerial head-quarters to the local level, are politicized one way or another and the political leadership often intervenes blatantly, making the officials only nominally non-partisan and non-political. Thus, it is not strange to see these officials seeking protection in rigid adherence to rules. The relationship varies enormously from country to country.

Administrative Structure

The newer approach – sanctioned by the World Bank, IMF, the US Agency for International Development (USAID), the British Overseas Development Agency (ODA) and other similar international aid agencies – appears to provide an alternative model to the panoply of problems associated with administered development, through 'de-administered' development. The opportunities for improper actions are lessened by reducing the scope of government activities: no bureaucratic structure equals no bureaucratic problems. Thus the market-friendly policies championed by the Reagan–Bush administration in the United States, by Brian Mulroney in Canada, or by Thatcher's Conservative government in the United Kingdom, as well as by numerous other Western European governments, are being universally applied. And since public administration is seen more as problem than solution, the granting of aid is made conditional upon policy changes involving a downsizing of the bureaucracy as well as the elimination of subsidies, the acceptance of devaluation, and other changes in monetary and fiscal policy. The first to be axed are the bloated para-statals, now to be entrusted to the private sector. At one time these were deemed administrative innovations that would circumvent the established rule-bound bureaucracy. Unfortunately, over the years, they have proved a bastion of patronage, often terrorised by powerful unions, and a drain on scarce resources.

Perhaps the fruits of these policies, as they are played out in the new Eastern Europe and the former Soviet Union (FSU), may become apparent by the time we enter into the twenty-first century. However, the record in the late 1980s gave considerable cause for concern: political resistance to rapid, large-scale sale of government enterprises; rebellion by government servants against privatisation; the scandal involving the Bank of Credit and Commerce International (BCCI); Southeast Asian and Latin American patterns of business ownership by top level government officials and their families, and the opportunities for graft when private interests gain control over public-sector operations; the power of enclaves such as drug cartels operating outside the normal governmental channels; and the coalitions formed between indigenous elites and multinational corporations. As Randall Baker has put it:

> The fear is that the state is handing over the national productive apparatus to the same internal privileged groups that it has tacitly supported, along with the foreign concerns that have the management and capital to take them over.[4]

Other new structural adjustments for development administration, such as of government downsizing and reorganisation, will be variously used as the twenty-first century approaches. Some of these will be indigenous, such as the small-scale industrial organisations in China, or public-works organisations such as Balinese water temples; others will be hybrids, cross-transfers from other developing countries, or western (northern) transplants. The alternatives to the traditional Weberian hierarchy are considerable; participatory 'bureaucratic populism' types[5] and community-based decentralised structures offer numerous options.

Administrative Personnel

If the downsizing of the bureaucracy continues, by the turn of the century there should be fewer public service officials proportionate to the population compared to the 1980s, even though some systems will continue to expand. Certainly, public servants at all levels and in all areas of the world should be better trained, more professionally oriented, more aware of the world at large, more ethical, more productive, and perhaps more humble. If this ever happens, morale will improve and the public attitude towards public servants will change for the better. In that case, administered development will co-exist happily alongside private sector entrepreneurship, as public officials provide a positive environment for economic expansion, unencumbered by red tape. I hope that this does not turn out to be merely another dream! Of course, if the economic environment of a nation warrants such a change, the downsizing may be easily accomplished.

If the new administrative environment emphasises public–private partnerships and market-friendly strategies, those managers who remain in strictly government service will need to reorient themselves towards a more facilitative role. This may open up new opportunities for abuse as existing informal techniques for avoiding and expediting myriad official regulations are supplemented by new rules aimed at making it easier for entrepreneurs to do business. The tendency to blame the bureaucrat for corrupt behaviour needs to be tempered by a realisation of how the whole system encourages unethical actions. Middlemen, touts, politician-protected elements, even the politicians themselves, play their role in the system. In Brazil the expediters are even unionized.[6] And nearly everywhere, public and private roles are easily mingled.

Administrative Procedures

Much attention is given by public administration specialists in the developed countries to systems approaches and management-information systems. These days, systems are computerised and administrators (especially in national and provincial capitals) are computer literate. In the developing countries, the process of computerising requires detailed analysis of operations. Properly done, the introduction of computers is based on a series of studies which focus attention on legal requirements, quality of statistics, overlap and duplication, cross-checking, and cost effectiveness. In India a national computer network has been established which encourages uniformity and interchangeability. It links districts, states, and the central government and provides information flows through both mainframes and microprocessors.[7] Numerous other examples of an intelligent use of computers may be cited, from Kenya's calculation of crop yields to India's use in railway-seat reservations, to Vietnam's computerisation training centres in Ho Chi Minh City.

Considerable attention by both indigenous and western specialists has been given to computerisation, with some important cautions regarding the pitfalls as well as opportunities in their use.[8] By the twenty-first century, the remarkable developments in telecommunications should make information available world-wide through networks using fiber-optic cables, digital switches, extremely powerful computers and methodologies of artificial intelligence. Those politico-administrative systems which are able to maximise the free flow of ideas and to promote the widest possible dispersion of information will be the ones which benefit.

This new era will produce a reduced role for the public sector and an expanded reliance on private enterprise; therefore numerous controlling and licensing activities will have to be eliminated or overhauled. This calls for streamlined decision-making, fair and uncomplicated rules and enforcement mechanisms, and opportunities to appeal against unfavourable decisions.

Towards a New Development Manager

This discussion has direct implications for the definition of both the tasks and the training of the administrative cadre.[9] A new scientific and technological system for the creation and extension of appropriate administrative know-how is needed. If development administrators are to be some sort of Schumpeterian entrepreneurs, a number of formative conditions along the lines of the paradigm outlined earlier must be met. In operational terms, this entails the necessity to combine techniques with training in those social sciences which analyse and act upon the environment.

Political and technical rationality forms the basis of this training. An administrator, from this perspective, is a link between specific experts (for example, agronomists, engineers, physicians), as well as between practitioners and the political sphere proper. In other words, a development administrator becomes both an operational analyst capable of problem-solving, and one equipped to implement policy through the integration of politics and technology. Furthermore, given the marked internationalised nature of present-day development and underdevelopment, such an administrator must go beyond an understanding of specific national realities. A general understanding of such topics as regional and world economics, natural resources and the environment, society, culture and politics becomes essential to the preparation of development administrators.

This does not mean that these individuals must be 'superprofessionals', displaying operational capabilities in politics, technologies and management proper. Once we recognize the complexity of the paradigm proposed here, in practice the training of administrators must proceed at different yet interconnected levels. Although all development administrators should recognize the different aspects of the analytical paradigm, the individual need not be competent in all of them. Three basic levels can be outlined here.

Technical Capabilities

The training in appropriate administrative technology should be oriented to two types of clientele. The first is grass-roots organisations. Training here would involve the expansion of simple technical know-how for management for change and problem-solving – a sort of 'barefoot managers'. This must be largely the task of extension work. The second clientele is essentially at the technical college and undergraduate level. In some cases, government and academic training, adult distance education and vocational training can play a significant role. Efforts in both sectors aim at the formation of a cadre of technicians able to see problems in perspective and apply basic management and organising skills to concrete and specific questions.

Tactical Analysts

Simultaneously, there is a need to prepare a fully-fledged professional – a tactical analyst – capable of understanding and acting in conjunction with the public sector. This middle management training requires a solid university education, with a social science background, so as to enable the practitioner to deal with the complex relations between administration and its context. This type of professional has been trained already in various quarters. However, at the present time, the emphasis has been on either an overly academic political-science training or a technocratic business-administration one.

Strategic Thinkers

It is also important to see the emergence of a third level: an academic cadre of 'strategic thinkers'. It is at this level where research and theorising in the field must take place. The role of this third level in the scientific and technological system described here is foundational. Strategic thinkers must elaborate a doctrine in the area of development administration which is adapted to local realities. Furthermore, they must serve as a centre for the diffusion of ideas and intellectual stimuli to the other levels.

In addition to these three levels, it is necessary to discuss the linkage between professionalisation and the improvement of the management-for-change capabilities. This relates to the exigencies of management training *per se*. Keeping in mind that grass-roots and rural management are by and large the first priority, the question here is what such training should contain. Obviously, a general answer for the Third World as a whole cannot be given. The training must vary in accordance with the specific circumstances of each country, region or programme. However, there are a number of common aspects which should be taken into consideration:

(1) The issue of improving managerial capabilities is not necessarily synonymous with the *professionalisation* of the public service. True, at the level of central agencies the two concepts seem congruent, but as soon as one looks at the myriad micro-organisations down the line to the grass roots, the ratios of appropriate management to professionalisation appear almost antinomic. For one thing, professionalisation is associated with a number of bureaucratic and transnational trends, which is just the opposite to the notion of need-oriented, self-reliant and autonomous development.

(2) *Managerial skill* involves much more than the sort of general administration which used to be the trademark of colonial administration. It requires a competence in the particular task at hand, whether it be in the field of agriculture, health, education, technology or industry. Management is neither an abstract operation nor a bundle of recipes. Disciplinary specificity must be considered when training programmes are conceived and put in place.

(3) There are some fundamental considerations of *content* that must be integrated into any curriculum. The desired goal, in its most general terms, is to produce competent, effective, responsible and ethical development administrators.[10] This entails the devolvement of a higher degree of substantive rationality, as mentioned earlier. Likewise, it involves the operational ability to solve concrete problems, particularly those related to the most basic needs of the population.

(4) The development of these skills also means improving *standards* and requisites of performance in handling developmental tasks. This includes setting high ethical standards and a public service ethos with the aim of providing a responsible and accountable development administration.[11] Public confidence and trust require that these managers maintain the best possible level of technical and moral performance, as well as a perceived dedication to achieving national as opposed to purely professional aims.

Transfer of management technology is crucial for the Third World nations, especially as they try to reduce their public bureaucracy under pressure from the North as well as from the international aid agencies. However, the most effective transfer in the managerial know-how will be when an indigenous, self-reliant capacity is created. This will be the challenge for both the North and South during the next few years and well into the twenty-first century.

CHALLENGES FACING DEVELOPMENT ADMINISTRATION IN THE TWENTY-FIRST CENTURY

A brief overview of some serious challenges facing development administration in the South in the next century is given below.

Growing political sophistication of the public. Over the years since the nationalist and independence periods, the public in developing nations has seen radical changes in the political and economic process; it has also come to realise that it does have some power to imprint its wishes. The public in those Third World countries where the democratic process is functioning can no longer be taken for granted by the politicians. A good example is the general election held in Kenya in December 1992. The public's sense of cynicism about the ability of those who govern and administer, particularly about the promises made by politicians, has made it more mature with respect to what to expect from those who govern. It now finds that more and more of these 'guardians' are selfish and self-centred; consequently, it is becoming hardened to the value of promises and the reliability of rhetoric. So, when it gets an opportunity, the public indicates its feelings through the weapon of the vote.

Growing influence of religion and traditional values in politics and administration. Religion has emerged once again as a force to be reckoned with in many of the Third World nations. In some cases religious fundamentalism has a mass appeal, especially when it involves a rejection of western-style modernisation, and replacing alien norms and customs with indigenous cultural and religious values. In many instances, these traditional values are at odds with western-style development, secular politics and steel-cold efficiency. While the administrative machinery continues to function within a framework of professed secular values, fundamentalism gets reflected not only in the political process but also in the style of administration. These tensions and conflicts should be appreciated so that the demands of fundamentalism and the role of religion are accommodated within the needs of development administration.

Political factors in managing the public sector. Development administration, sired by the West, assumed many western values, including the separation of politics from administration. It was thought that the management of the public sector would be done largely by administrators, ably assisted by economic plans formulated to achieve national goals. Consequently there was no place for the political dimension in the administration of development. However, as the Iran experience and similar instances elsewhere proved, politics could not be kept separate from economic planning, management of resources and administration of the public sector. Either the administrative system responded to political demands or politicians simply bypassed the established administrative apparatus and created their own network to accomplish their objectives. This created confusion and delays, which were blamed on the inability of leaders and administrators to run the machinery of government smoothly. Public sector management then became a source of economic ills, administrative mismanagement and political blundering. The separation of politics from administration remained artificially embedded in the parlance of theory and practice of development administration, while political factors dominated economic, social and administrative concerns. This realization is yet to be acknowledged by many scholars, practitioners and international aid personnel.

To reiterate, by their nature developmental issues are political, because they deal with the authoritative allocation of values in the context of limited and sometimes fast-diminishing resources. In developing countries, therefore, public sector management cannot remain purely within the domain of so-called value-free administration. Otherwise, irrespective of the amount of international aid, history may repeat the failed experiment of the USA's massive involvement in Iran. As a matter of fact, such repetitions have already occurred elsewhere. What is immediately needed is a new style of public sector management, which blends the political, economic, administrative, cultural and religious forces to produce the desired results.

Scarcity of new development managers. From the beginning of the era of development aid, the emphasis of both bilateral and multilateral aid programmes has been on economic planning, infrastructure building and the exploitation of natural resources. The training of development administrators received a low priority in the planning process. It was thought that once plans were set up in organic terms they would be self-administering. This neglect has been only recently acknowledged. In most of the Third World, development planning preceded the training and preparation of a cadre of administrators. Thus, as development projects multiplied, the gap between planning and management grew. The absence of appropriate managerial skills in the public service has led to either political or technical appointees who lack the managerial qualifications and experience for administering public programmes. In other words, the greatest institutional gap in the development process has been the underdevelopment of administrative and managerial skills.

As the Third World enters into the twenty-first century. The past few decades of developments in politics and administration have demonstrated two major trends: standards of conduct and probity have been steadily declining among politicians, although such a massive regression has not totally eclipsed the bureaucracy. Despite charges of overwhelming corruption, the administrative apparatus is functioning, and statecraft has not become completely contaminated by sectarian and similar insidious forces. And though the media continue to project a negative image of bureaucracy as being bloated, inefficient, status-conscious and authoritarian, that same bureaucracy is not totally cynical, bereft of idealism and dedication. The late 1990s are going to be crucial for Third World politicians and administrators, as they make efforts to uplift their people from the quagmire of poverty, malnutrition and underprivilege.

The challenge before the leaders and administrators of the Third World is, then, how to achieve the developmental objectives of basic human needs (the provision of food, appropriate habitat, health and education), as well as social justice and self-reliance, with very limited resources. They will have to be more self-reliant in the twenty-first century, as they cannot expect the same level of aid should the attention of the West turn towards helping Eastern Europe. So they must turn to their own resources among and between themselves much more than they have done so far. For this they will require a cadre of professionally trained and dedicated administrators, as well as moral and just politicians who can stand against the forces of corrupt politics and unscrupulous commercial and business interests. Such is the challenge and duty for the leaders and administrators of the Third World in the twenty-first century.

CONCLUSION

It seems that the administrative world of the twenty-first century – following the prescriptions of orthodox macro-economists and external aid providers – will be minimalist and facilitative. If there is a real-world model, it is Japan and, to a lesser extent, the four Asian dragons – Singapore, Hong Kong, Taiwan, and South Korea – along with possibly the newly industrialising countries (NICs) of Malaysia, Indonesia, Thailand and Mauritius, which may follow suit. One of the reasons for Japanese success may be its public bureaucracy which is 'one of the smallest and least expensive systems in the world. Yet it is an efficient, well-coordinated, and responsive public service.'[12]

Further, these Third World states need to bring back those conditions where a moral and accountable administration can flourish. Towards this end, every effort should be made to ensure that elected representatives and responsible public servants are held accountable for the proper conduct of their tasks. For, as I have stated elsewhere:

> The capacity of a nation's political system to prevent, detect, punish and control the abuse of power and authority will have a direct bearing on the legitimacy of a regime and will no doubt strengthen the moral basis of its government authority. Accountability for and responsibility in holding public offices can then be expected from our ministers and public servants. But, in the final analysis, accountability at the rank and file is the mirror of moral and responsible behaviour among the country's leaders.[13]

A strong role for government administration will be required as the next century approaches. Opportunities for corruption at the public/private interface must be guarded against without resurrecting redundant control agencies. The transformations in Eastern Europe and the former Soviet Union (EE/FSU) will constitute a test case, and we optimistically predict *relative* success in that region because of both the refocusing of western aid in that direction and the tradition of a relatively accountable bureaucracy. Third World nations are forewarned that businesses from the West are becoming less amenable to direct investments in developing countries because precedence is being given either to their own backyards (such as southern Europe), or to partnerships with EC countries or with other OECD nations. The tendency of European and North American industries to inward-directedness may become more pronounced once the last barriers to the free movement of capital have been removed. Of course, it is difficult to predict whether the overall effect of the Single European Market and the opening of the EE/FSU will be beneficial or adverse to the developing nations.

It is safest to assume that the overall conditions for developing nations' exports into the EC and North America will change rapidly as the 1990s come to a close. This will compel the developing nations to make substantive adjustments, although these may differ greatly from country to country. They will be hardest hit by new protectionist measures and other obstacles to imports in those single-markets. The high-performance NICs in East and Southeast Asia may perhaps adapt quickly to this new equation, but middle-range developing nations such as Egypt, India and Iran will encounter serious problems in keeping up with keener competition in the European and North American markets. For these developing nations it would be better to create some kind of southern regional marketing operation. For the poorer developing nations, however, the situation may not change much as they are going to remain heavily dependent, raw-material-exporting nations. Nevertheless, all Third World nations will have to come to grips with this new economic reality and the inward-looking emphasis of the West.

Thus, the prospects for the twenty-first century are not bright for the developing nations unless the West adjusts itself to recognise the difficulty of the Third World. Of course, it is easy for the free marketeers of the West to blame the Third World problem of over-sluggishly administered development on wrong policies and on a lack of openness to western ideas of economic development, the free market system, and use of modern technologies. The onslaught of these ideas will become even stronger through the 1990s. It will therefore behove the nations of the Third World to embark upon a kind of development which complements the winds of change brought about by the political changes in the EE/FSU region, and yet also *serves their specific needs appropriately*, as determined by their indigenous strategy for development. Any administered development will then be the midwife to deliver the goods.

We are living in a period of momentous changes which have far-reaching implications, particularly for the developing countries. The hitherto East–West divide, which some Third World states were able to manipulate in their favour, is a thing of the past. The collapse of centralised economic systems in EE/FSU has shown the viability of the western approach to economic development. It should come as no surprise to anyone in the Third World if economic restructuring is pushed further by the West or by the IMF/IBRD group of international financial institutions (IFIs). With such a push, restructuring will be imposed, and it will be particularly important in administered development as a key for strengthening institutions.

The developing nations have a vital stake in the orderly functioning of their administrative systems. The time has come for them to opt for accelerated development in order to make up progress thwarted by their creation of a thicket of rules, regulations, permits and the like in the past. The 1990s will test their capability to mobilise their financial and human resources. They can ill afford another decade of stagnation or arrested growth. Further, they should know that economic prosperity is not the monopoly of the West alone; any nation can aspire to and achieve it but not by remaining a bystander.

Notes

1. This section draws from Keith M. Henderson and O.P. Dwivedi, 'Administered Development: the Fifth Decade', *Indian Journal of Public Administration*, Vol. 38, No. 1, January–March 1992, pp. 1–16.
2. See J. Nef and O.P. Dwivedi, 'Development Theory and Administration: A Fence Around an Empty Lot?' *Indian Journal of Public Administration*, Vol. 27, No. 1, January–March 1981, pp. 42–66.
3. See Ralph Braibanti, 'Transnational Inducement of Administrative Reform: A Survey of Scope and Critiques of Issue', in John D. Montgomery and W.J. Siffin (eds), *Approaches to Development: Politics, Administration and Change* (New York: McGraw-Hill, 1960) pp. 133–83.
4. Randall Baker, 'The Role of the State and the Bureaucracy in Developing Countries Since World War II', in Ali Farazmand (ed.), *Handbook of Comparative and Development Public Administration* (New York: Marcel Dekker, 1991) p. 362.
5. John D. Montgomery, *Bureaucrats and People: Grassroots Participation in Third World Development* (Baltimore, MD: Johns Hopkins University Press, 1988).
6. Robin Theobald, *Corruption, Development and Underdevelopment* (New York: Macmillan, 1990) p. 157.
7. See Mukul Sanwal (ed.), *Microcomputers in Development Administration* (New York: McGraw-Hill, 1987).
8. See, for further reference, USAID, *Cutting Edge Technologies and Microcomputer Applications for Developing Countries: Report of an Ad-Hoc Panel on the Use of Microcomputers for Developing Countries* (Boulder:

Westview Press, 1989); OECD, *The Internationalization of Software and Computer Services* (Paris, 1989); Heinrich Reinesmann, *New Technologies & Management: Training the Public Service for Information Management*, (Brussels: International Institute of Administrative Sciences, 1987); William J. Stover, *Information Technology in the Third World* (Boulder, Co.: Westview Press, 1984); and Mukul Sanwal, 'An Implementation Strategy for Developing Countries', *International Review of Administrative Sciences*, Vol. 57, No. 2, June 1991, pp. 220–35.

9. This section is based on J. Nef and O.P. Dwivedi, 'Training for Development Management: Reflections on Social Know-how as a Scientific and Technological System', *Public Administration and Development*, Vol. 5, No. 3, 1985, pp. 245–7.

10. See O.P. Dwivedi, 'Ethics and Administrative Responsibility', *Indian Journal of Public Administration*, Vol. 29, 1983, pp. 504–17.

11. See O.P. Dwivedi and Nelson E. Paulias (eds.), *Ethics in Government: The Public Service of Papua New Guinea* (Boroko, PNG: Administrative College, 1984).

12. Paul S. Kim, *Japan's Civil Service System* (New York: Greenwood, 1988) p. 1. See also B.C. Koh, *Japan's Administrative Elite* (Berkeley: University of California Press, 1989). Koh credits Japan's civil service with maintaining a controlled environment based on carefully crafted policies which 'helped to minimize fallouts from adverse currents abroad while maximizing conditions for growth in strategic industries at home' (p. xi).

13. O.P. Dwivedi, 'Ethics and Values of Public Responsibility and Accountability', *International Review of Administrative Sciences*, Vol. 51, No. 1, 1985, p. 66.

3 Bureaucratic Morality, Corruption and Accountability

INTRODUCTION

'Bureaucratic morality' includes two terms: 'bureaucrat', which means a public official who is appointed, promoted, and retired or removed from the service through a merit system. Thus elected politicians and political appointees do not belong to this category. 'Morality' (or values) includes wider connotations than the term 'ethics', which has the narrower concern usually associated with unethical activities and codes of conduct. Morality includes both the positive and the negative values or attributes of holding a public office.

Ethics is often defined as a 'set of standards by which human actions are determined to be right or wrong'.[1] Ethical behaviour in the public service is considered as a blend of moral qualities and mental attitudes. The requisite moral qualities include not only the willingness to serve the public (or behave as a servant to the public) but also to behave competently, efficiently, honestly, loyally, responsibly, fairly and accountably. The mental attitudes include awareness of the moral dilemmas inherent in public policies which may still contain moral ambiguity, conflicting and competing claims by various community groups on the substantive and procedural aspects of administrative action, the empathy for divergent views held not only by some members of the public but also by colleagues in the service, and sensitivity for paradoxes of rules or procedures which may lead to frustrating and unkind actions. Hence, bureaucratic standards and requisites of performance should include what a public servant 'ought not to do', as well as an equal emphasis on the positive side of 'ought to do.' Rectitude and 'duty to serve the public' are both parts of the expected standards of the public servant, as is an acknowledgment of the moral implications of public policy and administrative actions.

It should be noted that there is a difference between a 'legal' action and the 'right' action. 'Doing right' requires more than a legally justified action; it means looking at both sides of the case so that one can make a fair decision, which may require going beyond the legality of a case. Many a public servant will not venture beyond the legal requirement of a case irrespective of the moral implications or the problem of fairness. Of course, the public expectation about bureaucratic morality is high, akin to what it expects from clergy, nurses, or parents. The public still believes that those who work for the state are there primarily to honour the call to serve the nation. Consequently, when public servants go on strike to demand better working conditions or higher wages, the public is generally unwilling to change its perception. And whenever there is a report in the media about some unethical activity by a public official, the public reacts as if the event should not have occurred, particularly in government.

A change has been noticed in the attitudes of recently recruited public servants. They now view their employment as a 'job' rather than a 'vocation'. They decline to accept the traditional roles of public servants wherein duty, patriotism, serving the community at all times, and sacrificing the personal life for higher values are considered more important than the industrial, unionized approach to work ethics. Thus, a conflict exists between the traditional public expectation from those who serve the state, and the modern employees of the state who believe that their obligations to the state are restricted to those duties which are part of their job classifications, and for which they can be held legally, but not necessarily morally, responsible. Thus, in recent years the meaning of professional standards has acquired fewer moral and more legal connotations.

This chapter discusses the various aspects of bureaucratic morality in the developing nations and the issue of administrative accountability. In the process I hope to demonstrate, through various examples, what kinds of ethical problems are confronted by public officials in different settings and the various attempts of their governments to tackle these problems. This reflects a common concern, one that is central to the field of public and development administration in all cultures: how to raise the moral consciousness of public servants and ensure that they behave morally, responsibly and professionally.

ADMINISTRATIVE CULTURE AND MORALITY

For many developing countries, the decades since independence have
not shaken off the basic foundations of the imperial heritage, particularly with respect to the administrative machinery. There is no doubt that
the foundation is not as solid and functional as it used to be during the
early years of independence; indeed, it has received a continuous
onslaught from both developmental imperatives and the traditional culture. Nevertheless, the outward structure has survived. The colonial
powers were able to maintain their control with only a handful of
administrators for more than a century. The independent era could not
retain most of the positive elements of these colonial administrations
(such as dedication, political neutrality, and impartiality); instead there
has been a continuation of many of the negative traits (such as emphasis
on centralisation of power, authoritarianism, a tendency to check abuses
rather than take positive action, a distrust of the business and commercial sector, an exaggerated concern about status and privileges, and
arrogant attitudes towards the common man).

Compounding the above has been the emergence of a new breed of
political leader in these countries, for whom the main tenets of a west-
ern-oriented secular and politically neutral bureaucracy have been an
irritant. These new political leaders also found that such inherited
administrative values slowed down the pace of social and economic
reforms. Thus, a conflict developed between political expediency in
achieving national objectives and the professionalism of public servants. However, it was evident that the politicians would emerge as the
most important actors. Ministers claiming the right to shape the destiny
of the nation told the public servants that less rigour must be placed on
the inherited colonial values of professionalism, neutrality and objectivity if nation-building so required. Of course, no one questioned the
primeval right of the politicians. These people had, after all, fought for
independence while the public servants had continued to work for the
colonial masters. It was not surprising then to see power shifting from
administrators to politicians as the latter acquired supremacy in decision-
making matters, including the control of public servants. This shift is
akin to the western concept: keep the public service on tap, not on top.

With this shift in the power base, particularly in civil service
appointments, promotions and the use of discretionary decision-
making authority, certain side-effects were inevitable. For example,

politicians were found acting as brokers between business concerns and government departments; the interpretation and enforcement of laws was politicised; interference in normal personnel administration to secure the appointments of friends or supporters, or the promotion of local civil servants, became commonplace; the sale of government property and the issuing of contracts and licences came under political influence; police, para-military and military forces were used improperly in the normal functioning of society's affairs; there was political manipulation and intervention in the purchase of machinery, property, equipment and services for government departments; official and confidential information was misused by politicians for private gain; and the concentration of extra-constitutional and legal authority was placed in the hands of favoured individuals who did not hold any elected position. These activities have been supported by an emerging but powerful economic class which has sought immediate personal gain, and politicians have become an easy conduit for achieving such ends. Naturally, such an environment influences the conduct and attitudes of public servants who, by and large, see the benefit of adjusting to the situation.[2]

In that environment, bureaucracy attempts to shield itself by weaving an intricate array of rules and regulations, which in turn results in inadequate delegation of responsibility, the requirement of excessive concurrences, and a general distrust of the public. One example of administrative distrust against the community of businessmen is a highly controlled system of corporate licensing, monitoring and regulation of all kinds of commercial and industrial activities, and the general surveillance of the private sector. This has led to an obsessive preoccupation among administrators with disbelieving whatever the business sector does or says and has given rise to intense animosity between the administrator and the entrepreneur. For the entrepreneur, corruption then becomes the only means by which to secure administrators' favours or their silence. In these circumstances, a new trait of administrative culture in many developing nations is discernible: the existence of a parallel 'black administration' where influence, favours, money, privileges, misuse of public funds, falsifying records and bending of the rules play the crucial role. Complementing this is another trait: being excessively meek before superiors but revelling in bullying and bossing subordinates (and sometimes those members of the public who come to seek official favours).

The public believes that corruption and unethical activities among elected politicians and public servants have become so widespread and deep-rooted that it is futile to worry about it. Their pessimism is justified because they know that the custodians of government are in the unholy grip of corruption (and corrupt demands of politicians), and nothing moves smoothly without the help of influence, connections, or bribery. For example, cases of non-existent or fictitious persons receiving bank loans have been reported by the central Vigilance Commission of India. So the fundamental question emerges: Is there a limit to official corruption? Sadly, no one has been able to suggest any such limit or a level of toleration. Nevertheless, there is a need to protect a small band of honest and dedicated public servants who are battered by politicians and power brokers, and who need the support of the public they seek to serve.

While the extent and scope of institutionalised bureaucratic or political corruption may not be so pervasive as to affect the whole administrative machinery, the public views it so. One way to restore confidence would be to take swift and harsh action against those political leaders and public servants who have been charged with unethical activities. While it is easy to transfer, demote or even dismiss a public servant so charged and found guilty, it has been the tendency not to press the charges against political leaders if they belong to the party in power; and even if they are forced to resign they are often reinstated as ministers after re-election. When such instances become a regular feature of the governing process of a country, the moral fibre of the society declines. Consequently, everyone joins in the game of exploiting public office for private gain. Naturally, some public servants, not wishing to be left behind, eagerly join the race. Then everyone laments the sorry state of government and expresses helplessness in the face of moral decay.

THE CULTURAL CONTEXT OF BUREAUCRATIC MORALITY

The bureaucracy of a developing nation suffers from certain strange paradoxes. A rigid adherence to procedure is combined with a ready susceptibility to personal pressure and intervention. While a bureaucrat may give the appearance of being preoccupied with correctness and propriety, in practise s/he may be committing endless irregularities and improprieties. Similarly, his or her apparent pursuit of the uniform application of absolute justice may contain glaring anomalies. It is a curious reflection on their attitudes and thinking that these bureaucrats are willing to tolerate such contradictions between theory and practice.

In several countries, the civil service is characterised by excessive self-importance, indifference towards the feelings or the convenience of individuals, and by an obsession with the binding and inflexible authority of departmental decisions, precedents, arrangements or forms, regardless of how badly or with what injustice they may work in individual cases. Additionally, the civil service has a preoccupation with activities of particular units of administration and an inability to consider the government as a whole. The civil servant also fails to recognize the relationship between the governors and the governed as an essential part of the democratic process. There is no appreciation of the citizen's viewpoint, and exercises in public relations are aimed more at publicity and propaganda than at establishing rapport with the community or making genuine attempts to involve the public. Some bureaucrats, when confronted with a difficult decision, seldom make any attempt to tackle the problem with initiative and imagination; they merely refer the matter to another department or make a series of unnecessary references to subordinates to gain time. It is not difficult to discover the reason for this: after all, non-performance has never been regarded as grounds for disciplinary action. The result is a psychology of evasion wherever possible. There are even year-long delays in the implementation of major decisions.

Other maladies often attributed to the Third World bureaucracy stem from its structural characteristics. The entire machinery of administration today is groaning under its own size. The proliferation of schemes, however, is not commensurate with the availability of qualified persons. While line jobs remain understaffed, staff positions, which have an irresistible appeal for the action-shy and the comfort-loving, have increased exponentially. Although there is a core of exceptionally hard-working, dedicated and conscientious officers, they are today overwhelmingly outnumbered by the complacent, who are obsessed with status, rank and emoluments, and addicted to habits of personal luxury and indolence.

In sheer desperation, many scholars tend to argue that the differences in efficiency and productivity between the Third World bureaucracy and its counterpart in the West are the inevitable results of differences in national character. If we accept this explanation, we are in fact left with no likelihood of remedial action. On the other hand, if we look towards the private sector of these nations, instances of efficiency and great productivity do abound. Obviously, there ought to be other explanations for the phenomenon. Is it because of the structural characteristics of developmental bureaucracy, which possess an unmistakable resemblance to the traditional culture and the milieu in which they operate?

Complex as it is, Third World culture is itself a product of the influences of flourishing folk cultures, the sub-cultures of tribes, castes and classes, and the urban culture of an English-educated middle class. It is an amalgam of various traits, traditions, attitudes and outlooks that draw on the traditions and religious background of these societies. An official's behaviour will, therefore, also be culturally determined. In spite of his or her participation in a depersonalised system, the official's behaviour must necessarily be culture-bound or s/he risks social disapproval. I commented on this aspect of the problem some time ago:

> An administrative system influenced by such traditional loyalties will tend toward an ascriptive rather than achievement-oriented pattern of recruitment. And that is why a person who asks favours from officers belonging to his caste does not consider his act unethical. Similarly, when a government official 'fixes' applications and licences in utter disregard to merit but in accordance with family and caste loyalties he is obeying a law of social conduct more ancient than that of the upstart state. Moreover, in any traditional society the family forms the common interest group 'par excellence'. ... Since family in this usage includes uncles, cousins, nephews, grandparents and other near and distant relatives, a great deal of pressure to 'fix' jobs for them or to find some other source of income to support them is not uncommon. ... Moreover, these relatives ... would like to take advantage of his position in securing jobs, procuring permits, etc., while he is still influential. They do not consider the exploitation of a relative's official status as something bad or unethical.[3]

The orientation of officials in several of the developing nations is not merely based upon personal economic and social conditions, but also on what is often termed 'ethno-expansionism'. This involves the assumption that one's own ways are superior to all other ways and that the circumstances one enjoys should be desired by all people. The concept of ethno-expansionism is of particular value to the analysis of the cultural content of developmental bureaucracy. Thus, culture plays an important role as a modifier and attitudinal change agent.

BUREAUCRATIC CORRUPTION: A COLONIAL LEGACY?

Some observers have argued that bureaucratic corruption in developing countries is merely a logical evolution of the acceptance of western behavioural patterns. Colonial people discovered that western norms were not applied universally by their colonial masters: there was one rule for Europeans and another for natives. Corruption was abhorred if Europeans were involved but was accepted as the natives' way of life. The British Indian Civil Service officers, for instance, had separate clubs and a general tendency to exclusiveness from the people whom they were ruling. Exclusiveness was essential in order to maintain impartiality and the ability to disengage. To most colonial administrators nepotism, bribery, the institution of polygamy and the publicly condoned acts of cruelty all suggested inferiority of race and norms. And it was perhaps because of this that they became apathetic and cynical toward corrupt practices prevalent in colonial society, and did little to control such unethical behaviour among natives, even though at the same time they were not permitted among themselves. The result was that some forms of corruption became institutionalised and were carried over even after independence.

It also became apparent to colonial people that merit and achievement criteria were disregarded by their masters when the question of religious affiliation was considered. Neo-converts into Christianity were given preferences over qualified non-Christians in civil service appointments and promotions. In Africa, for example, people drawn from the lowest stratum of traditional society were elevated, through the activities of missionaries, to the highest strata of the social scale. Similarly in India, Anglo-Indians were recruited in several government departments largely on an ascriptive basis. Today, for example, when the widespread ascriptive considerations in the appointment of government employees in India are criticised, people tend to refer back to the colonial recruitment system to rationalise the present situation. In the developing nations primary associations are still dominant: family, kinship, caste, neighbourhood, village, ethnic origin, and religious affiliations are the associational forms that have the first and the greatest call on individual loyalties. There are modernising elites in each developing country but they too are swayed by such ties. Most civil servants, though possibly recruited on the basis of merit and competition, maintain their traditional ties.

In a traditional society a bureaucrat faces two sets of values. Trained in western norms, he or she publicly adheres to the norms of objective and achievement-oriented standards for recruitment and selection. But privately, he or she is forced to subscribe to a rigid hierarchy of caste, tribal affiliation and particularistic norms. Thus it should not surprise a Westerner visiting a developing country to find civil servants publicly condemning bribery and corruption, but at the same time secretly indulging in it in the name of helping their kith and kin. Is the existence of such a double-standard due to the identity crisis faced by civil servants?

Other factors which may cause public servants to resort to corruption are job scarcity, an inadequate salary, and the ever-increasing powers being given to them by the state to regulate its economy and social affairs. This increased regulatory authority creates various opportunities for money-making: for instance, in connection with planning permits, contracts for construction, import–export licences, collection of customs and other duties, and strict accounting for foreign exchange.

To summarise then, corruption can exist only if there is someone willing to corrupt and someone capable of corruption. The materialistic way of life has a tendency to subvert the integrity of those who are in the public sector. However, those at the top of the political and administrative system must set the tone for an entire public service. It is strange that the public often demands a high standard of morality from government employees but at the same time condones unethical practices in the private sector, or among themselves.

THE IMPACT OF POLITICAL CORRUPTION

In Third World nations, the incidence of political corruption has assumed frightening proportions. It has not only spread to nearly every part of the governmental machinery, but has also grown even more rapidly amongst the professional politicians, the party people at all levels of the state and the central governments. The peoples' perception of political corruption as a fact of life is total and documented daily. Today corruption is not the despised word it used to be; on the contrary, it is considered as a part of normal behaviour.

In the past, especially immediately after a nation's independence, corruption was seen as a venality. Of course, there had been corruption previously, and people whispered about it, but it was hidden and was perceived by the public as the worst crime a politician could be found committing. But when political parties needed resources to continue their hold on the government structure, selling favours and interfering in the normal administrative process became a routine practice. At first the media and people in opposition exposed several corruption cases, but the various changes in governments and leaders produced no result. It became expected that anyone who had power would feel obliged to use it for his private gain or for the assistance of his friends and relatives. The situation changed radically: initially leaders suffered hardships during the struggle for independence, and later there was a new breed of politicians assisted by administrators and commercial interests whose selfishness seemed to have no limit. This was further compounded when a nexus appeared between criminals and politicians, especially at the time of general elections. A new ethos has appeared which seems to say that one has the right to go to the devil in whatever way one chooses in order to secure betterment for self and relatives. The fear of religion, which would have restrained people from becoming cheats, is no longer there because even the religious institutions are not immune from problems. The politicians of today have developed their special ways of securing funds from others, and in this process they are ably assisted by some administrators. This means that the poor have only one way to remove their discontentedness – by violence. People are becoming brutally violent and there is a rise in criminality. Is there a limit to it?

Political corruption in these nations has assumed at least two new dimensions:

(1) *The exercise of extra-constitutional authority*: The first is the emergence of extra-constitutional centres of power which exercise enormous influence and authority on behalf of the legitimately constituted institutions (e.g. Sanjay Gandhi during his mother's rule). The irony is that while the conduct of public office-holders in many cases had ostensibly remained above-board, their sons and daughters or close relatives have amassed huge wealth, power and status through the exercise of undue influence. The politicians have invariably protected the interests of their relations and protégés, pretended ignorance of any allegations of corruption on their part,

or risen to their public defence when challenged by the press or the opposition. However, the politicians' abuse of authority, and their interference with the process of governance on behalf of ambitious wives and other relatives, continues to this day.

(2) *The rise of the professional politician*: A second form of political corruption, which has made inroads into the body politic, comes from a new breed of politician known as party fundraisers or bagmen. Elections have become an expensive business, and consequently the emphasis in each party has shifted from issues to fund-raising by any means possible. When a party is elected to office, these clever and energetic fund-raisers are often rewarded with the economic development ministries, which issue the largest contracts and the most important licences and permits. The technique of fund-raising for political purposes has reached a new height of influence-peddling and interference in the regular system of administration. The distinction between this new kind of politician and those of the older generation is that while the latter often used their power arbitrarily, their chief aim was to advance the public interest and achieve public goals; the new politicians have no qualms about circumventing established norms in order to raise funds for their party. In such a situation, bureaucratic morality is overridden by political expedience.

AN ACCOUNTABILITY PERSPECTIVE

Accountability as a term in the theory and practice of public policy and administration has been used often; traditionally, it has meant an answerability for one's actions or behaviour. Generally, public officials and their organisations are considered accountable only to the extent that they are legally required to answer for their actions. In this narrow sense, legality of administrative action acquires prime importance. However, the public does not perceive that the accountability of public officials and their agencies is limited only to the legality of their actions. From the public's perspective, other aspects of accountability, such as organisational and professional behaviour, political elements, and morality of administrative actions, are equally important factors. Thus the term 'accountability' in the context of public policy and

adminstration ought to include at least five characteristics: organisational or administrative, legal, political, professional, and moral. However, before we discuss these, a more precise definition of this term is
necessary.

Perhaps a broadly conceived definition could be constructed as fol-
lows: public service accountability involves (a) the methods by which a
public agency or a public official fulfils its duties and obligations, and
(b) the process by which that agency or the public official is required to
account for such actions.[4] Viewed as a strategy to secure compliance of
entrusted duties and a means to minimise the abuse of power and
authority, public service accountability acquires a variety of elements
which are briefly discussed below.

Administrative/Organizational Accountability

An agency's or organisation's accountability requires a clear cut hierarchical relationship between centres of responsibility and units where such
commands are acted upon. The hierarchical relationship is generally and
clearly demarcated, either in the form of formally pronounced organisational rules, or in the form of an informal network of relationships. Priorities as determined at the superior level are followed, and supervisory
control is exercised intensively with a clear understanding of the need to
follow orders. Any disregard of such orders is harshly dealt with, the
punishment ranging from informal reprimand to eventual dismissal.
However, employees are able to escape from 'organisational brutality' if
they can prove that the orders were not clear enough or if the resources
required to carry them out were inadequate. With the growing influence
of unionism, it has become difficult in any public service organisation
clearly to pinpoint the perpetrator of the error. This was amply demonstrated in the space shuttle Challenger's explosion on 28 January 1986.[5]

Legal Accountability

While bureaucratic accountability relies on internal means of control,
legal accountability seeks answerability of an action in the public
domain through the established legislative and judicial process. This
could be achieved by a court action, or a judicial review of an administrative action whereby the organisation or its officials are held accountable for not following the legislative intent or the legal obligations so
required. While legislative or judicial power to punish the administra-

tion is neither swift nor extensive, legal accountability does get applied, sooner or later, or the law gets changed.

Political Accountability

This relates to the issue of the legitimacy of a public programme, and eventually the survival of that organisation which is entrusted with a particular responsibility. In all democratic systems of government, administrators are duty bound to recognise the power of political authority to regulate, set priorities, redistribute resources and ensure compliance with orders. Political accountability, in many cases, subsumes administrative or organisational accountability, mainly because the elected politicians assume the responsibility for getting a job done and the necessary results produced. Sometimes this results in the politicisation of administrative accountability. In such a situation, a confused pattern of accountability emerges where politicians may be eager to accept credit but reluctant to acknowledge blame, and instead deflect such responsibility on to administrators. Political expediency, in shifting accountability, then takes over the political system.

Professional Accountability

With the advent of professionalism in the public sector, there have been demands by professionally trained public servants (such as medical doctors, engineers, lawyers, economists, social workers, accountants and the like) that they be trusted to do the best job possible. They expect to receive sufficient latitude in performing their tasks, and in determining the public interest; and if they are unable to accomplish their jobs, they expect some reprimand. But as modern governments are becoming more and more dependent on professionals for their expert advice, the nature of public interest as defined by these professionals does not mirror necessarily the composition of society. Consequently, their advice, and thereby the public policy which is based on such advice, becomes skewed. For example, if medical doctors are to determine the fees to be charged to the public in a medicare programme, then their behaviour can be seen at best as 'self-enlightened interest'. Professional accountability demands that professionals in the public service should balance the code of their professions with the larger objective of protecting the public interest. Sometimes these two aspects do not coincide, and sometimes there are parallel or competing demands. But ultimately, it is the public interest which should determine a professional's responsibility and accountability.

Moral Accountability

It is now widely accepted that governments ought to be legally and morally responsible for their actions. The state, as Hegel pointed out, is a moral organism, which means that it is more than a system of legal or organisational norms. The activities of a public official have to be rooted in moral and ethical principles as acknowledged through constitutional and legal documents, and accepted by the public through established societal norms and behaviour. And since all governmental actions become moral actions, it is normal for the public to expect moral conduct from its politicians and appointed public officials. However, there is a growing cynicism that one cannot trust those who are in power and authority. Increased instances of political and bureaucratic corruption have added further fuel to this already burning issue in the public domain. Hence, the demand for moral accountability in the state has been increasing all over the world. A moral public official is not simply one who obeys the laws and behaves within the confines of bureaucratic norms,

> but also one who strives for a moral government. Such is the duty for those who wish to be involved in the difficult and complex world of statecraft. This is the essence and basis of a moral state. Only by demonstrating the highest standards of personal integrity, honesty, fairness, justice and by considering their work as a vocation, can public officials inspire public confidence and trust, the true hallmarks of a moral government.[6]

Moral accountability, which subsumes administrative, legal, political and professional dimensions, will then be possible. And only then can one hope to secure a moral government.

An Overview

The foregoing review of public service accountability leads to the conclusion that appropriate mechanisms for securing responsible and accountable government are still developing. And yet account-

ability as a viable instrument of control can be successful only to
the extent

(1) that public servants understand and accept their assigned
    responsibility for the results expected of them;
(2) that they are given authority commensurate with their
    responsibility;
(3) that acceptable and effective measures of performance evaluation
    are utilised and results communicated both to superiors and to the
    individuals concerned;
(4) that appropriate and equitable measures are taken in response to
    results achieved and the manner in which they are achieved;
(5) that political leaders and ministers make a commitment to honour
    these accountability mechanisms and procedures and refrain from
    using their positions of authority to influence the normal func-
    tioning of the administration.[7]

How can a moral and accountable administration be achieved?
First, what is needed is a change in the attitudes and behaviour of the
public. Unethical conduct in government is shaped and conditioned
by such behaviour in society. 'Therefore, attempts to reform public
bureaucracies as independent systems are necessary but not sufficient'.[8] The public must show courage and if need be accept sacri-
fices to oppose such practices. This is possible if people believe in an
ideology and are willing to suffer for it, or if, as a result of a revival
of religious values, they become so disgusted with the existing situation that they are willing to take drastic action against the offenders.
Nevertheless, every effort should be made to ensure that elected representatives and responsible public servants are held accountable for
the proper conduct of their tasks. Of course, the capacity of a
nation's political system to prevent, detect, punish and control the
abuse of power and authority will have a direct bearing on the legitimacy of a regime and will no doubt strengthen the moral basis of its
governing authority. Accountability for and responsibility in holding
public offices can then be expected from our ministers and public
servants. But, in the final analysis, accountability in the rank and file
is the mirror of moral and responsible behaviour among the
country's leaders.

THE MORAL DIMENSION OF PUBLIC-SECTOR MANAGEMENT

The contemporary interest in the moral dimension of government is attributable to several factors: the continued growth in size, scope and complexity of government and its resultant negative attributes (generally referred to as an overpowering Leviathan in the form of the administrative state); the insistence of the public on open and accountable government; demands for enhancing and protecting individual rights and freedoms; a general feeling of disappointment with the conduct of elected public officials, and frustration with the erosion of the concept of service and the doctrine of vocation among public servants; a growing cynicism about the capacity of government leaders to protect the quality of the environment, and in their ability to enhance human dignity; and finally, a deep feeling that people in politics and administration are not to be trusted. These and related factors have increased the demand for more moral and accountable government.

However, the contemporary moral outrage expressed world-wide by the public in the 1970s and 1980s differs from the demands for civil service reforms in the West during the late nineteenth century and early twentieth, when the emphasis was on transforming the civil service from a corruption-riddled nest of patronage into a professional service. Those earlier reforms, coupled with the growth of the administrative state, gave immense discretionary power to public officials.[9] The use and abuse of that power in recent times has created demoralising and dehumanising tendencies in public bureaucracies. The growing public outrage is a reflection of such tendencies in government and administration.

At the heart of the morality issue is the people's mistrust of the power of government and its possible abuse. Any instance of unethical activity by a public official adds to the growing loss of faith in the fairness and objectivity of both elected and appointed public servants. To the public, examples of patronage appointments, organisational brutality, the use of power either to obstruct or delay the desired action, conflicts of interest and unethical conduct are the symptoms of that widespread malady. These observations are especially relevant to situations where opportunities may arise for administrative excesses and the arbitrary use of power. Safeguards, checks and balances must be instituted, whereby the actions of public servants are scrutinised regularly and openly, and where any infraction is immediately and appropriately dealt with. If the administrative apparatus is to serve the people, the public officials must be accountable, not only for the services rendered, but also for the manner in which they are delivered.

Some important questions, however, arise from this assertion. Are public servants aware of the ethical implications of the power they exercise? What values should they choose in the exercise of their responsibilities?[10] Their choices will clearly vary, depending in large part on their conscience, their family and religious background, their country, their government, and the particular department in which they are employed. But, in the absence of any agreed values, public servants are left in a void which increases the ethical predicaments they face in the exercise of discretionary power and the management of public programmes.

Obviously, there is an urgent need to develop the public's awareness of the abuse of power and authority by public officials in some countries, as there is a need for constant vigilance in those nations that have achieved responsible and accountable public services. Consequently, the prime goal of any administration ought to be to ensure that both elected representatives and appointed public servants are held responsible for the proper exercise of power. The abuse of power in the public sector undermines public confidence and trust in government, reduces the capacity of government to fulfil its functions effectively, subverts ethical responsiveness to the citizenry, and imposes an unnecessary financial burden on the public. The capacity of a political system to prevent, detect, punish and control such abuses has a direct bearing on its legitimacy and the moral basis of its authority.[11]

ADMINISTRATIVE THEOLOGY

What can be done to recapture that sense of mission, dedication and service which used to be the hallmarks of our public officials? In this section, the concept of the 'theology of administration' (or administrative theology) is discussed as a possible source of encouraging that mission and dedication towards a moral government.[12]

The theology of administration may be seen as a study of concepts and practices relating to matters of ultimate concern in statecraft. Administrative theology, on the one hand, subsumes administrative ethics which 'involves the application of moral principles to the conduct of officials in organizations'.[13] Administrative ethics, in turn, is related to (and is generally influenced by) political ethics.

> Administrative theology, on the other hand, drawing on the doctrine of vocation (or callings) and the concept of service, relates to the sense of mission which a public official is supposed to undertake, to serve the public, perform duties and fulfil obligations.[14]

The objective of this concept is based on three specific elements.

First, the term 'theology' is used here in a most ecumenical way. As most modern states exhibit multi-cultural diversity, administrative theology can strengthen the moral dimension of government by drawing at least one special feature from all high religions of the world. That feature is the concept of *service*. In all religions, people have been exhorted to serve others; in each religion as well as in all cultures, this doctrine is considered to be the ultimate concern of all human beings. Thus, when such a concern is expressed through a vocation such as public administration, whose ultimate aim is to serve humanity and to protect the public good, it acquires the desired prerequisite of moral government.

Secondly, the term draws on the doctrine of *vocation*, which is evident in all major religions of the world. For example, a similar doctrine of vocation was prescribed several thousands years ago by Lord Krishna in *Bhagavad Gita*: 'One must perform his prescribed duties as a vocation, keeping in sight the public good'.[15] The doctrine as it relates to statecraft, and specifically to the duties and obligations of public officials, is still relevant and requires urgent revival.

Thirdly, the use of the term 'administrative theology' could relate to different interpretations of the established *norms* of modern secular bureaucracy. For example, the concepts of neutrality and objectivity in this context would not mean total disinterest and non-involvement as well as avoidance of morality; rather they would require the sympathetic involvement of public officials in protecting, enhancing and serving the public good. It also means displacing or minimising such traits of the administrative state as impersonality, authoritarianism, red tape and bureaucratic oppression, by believing in the vocation of public service as a true calling to be a servant of mankind. All religions agree that the true measure of a religious man or woman is whether he or she is able to serve others.

These three major elements suggest that administrative theology can be an important dimension of moral government and administration. Religious precepts can be a formidable force to support any democratic form of government as long as we are not compelled to accept any theocracy or orthodoxy that suppresses the acknowledgment of other viewpoints, leads to a uni-directional handling of public policy and administration, or suppresses existing democratic values.

Is administrative theology possible? The two most serious obstacles come from the foundations of the modern state and administration – (1) the separation of the state from religion, and (2) the presumption of the neutrality of administrative action. Neither of these needs to be displaced in order to include administrative theology as one of the basic standards and requisites of performance of administrators; also, by introducing such an element, we might find ourselves in a self-consciously moral society which would have to put duties and obligations first, and relegate individualism and selfishness to a subordinate place in the democracy.

The moral foundation of any public service organisation in a democracy requires that administrators and public officials show a genuine care for their fellow citizens. Devoid of such a moral foundation, a situation could emerge akin to the Nazi bureaucracy, when state administration was enlisted in the cause of evil, led by self-righteous people in government who sacrificed the moral obligation of the profession of serving the public.[16] Democratic values such as equality, law, justice, right and freedom have moral connotations, and demand an unwavering commitment to serve from those who govern. Public officials are obligated to uphold these values (which may be enshrined in a nation's constitution or considered as self-evident truths), because in the final analysis the appropriate implementation and enforcement of government policies and programmes rests upon their shoulders. This is the basic premise upon which depends the normal functioning and survival of a democratic system of government. Public administrators have an obligation to serve the public in a manner which strengthens the integrity and processes of a democratic society. This is a resolve which draws on the doctrine of vocation as it relates to administrative theology.

The essence of this concept is in the identification of one of the strongest assets in any individual; that is, serving others. Its basic thrust is to motivate people in government so that they can make a full contribution of their capabilities in better serving their country and the public:

The plea is addressed mainly to those who are the backbone of the statecraft – public servants – and the objective is to create an environment, an administrative culture, in government so that public servants as well as ministers are able to respond to the challenge of moral government. But such an environment will not result by itself unless there is a change in the management philosophy, attitudes of public service unions, and conduct of elected and appointed public servants – all to be oriented towards the broader aims of this plea: achieving excellence by serving others.[17]

Public Service as a Vocation: The Ultimate Value of Administrative Culture

Morality has been a guiding force in the history of mankind, particularly in statecraft; it informs how we are governed, our relationship with others (individually and collectively), and our understanding of the nature and destiny of man. While the emphasis on secular government and democracy may have relegated the place of morality to individual conduct and behaviour, it has nevertheless maintained a continuing tension between the requisites of public policy and programmes, and the moral standards by which they can be measured. But that tension has not been maintained as we observe that immoral and unethical activities in public places are on the rise all over the world. If justice, equality, equity and freedom are to be maintained, proximate political and administrative acts must have moral standards by which they can be judged. As I have advocated elsewhere:

> All governmental acts, if they are to serve the present and future generations well, must be measured against some higher law. That law cannot be a secular law because it is limited in vision as it is framed by imperfect people in their limited capacities. That law has to be, perforce, based on the principles of higher spiritual and philosophical foundations. Administrative theology is one such foundation which can provide an important base to a moral and responsible statecraft.[18]

The ideal of administrative theology draws upon the concept of self-sacrifice – a concept which rises above individualism and hedonism or materialism in order to create an environment or spirit of public duty among government officials and bureaucrats. This does not mean that public servants must take a vow of poverty: rather it means adhering to the principle of serving others by setting a high standard of moral conduct and by considering their jobs as a vocation. I appreciate the difficulty in realising this ideal, particularly when individualism and unionism have become the dominant forces moulding the administrative culture; nevertheless, many public servants will rise to the challenge when confronted with the choice between some minor discomfort to their own well-being and the inner satisfaction to be gained from rendering service unto others.

However, administrative theology is not to be equated with bureaucratic ideology or even theocracy, because it is reconciled with the higher values of democratic secularism and morality. Further, the morality which determines political and administrative action is multi-dimensional. It is rooted in the civilisation of mankind, and derives from its nobler foundations. It draws from the community of nations and various cultures, and influences the universe in which we live. Confidence and trust in democracy can be safeguarded only when the governing process exhibits a higher moral tone, deriving from the breadth of moral dimensions. This calls for a commitment on the part of elected and appointed public officials towards a moral government and administration. Actually, we get a moral government by creating those conditions within which such a government can operate by making it possible for officials to acquire the necessary traits, and by the practice of the same. A moral public servant, to be specific, is not simply one who obeys the laws and behaves within the confines of bureaucratic values, but one who strives for a moral government. Such is the duty of those who wish to be involved in the difficult and complex world of governmental bureaucracy. This is the essence and basis of a new administrative state where public officials are able to inspire public confidence and trust by demonstrating the highest standards of personal integrity and morality.

EMERGING CHALLENGES

Three major challenges influencing the domain of public sector ethics are discernible: how to regain spiritual guidance for secular affairs; how to combat growing scepticism about government credibility; and how to move towards a moral administration. These are briefly discussed below.

Spiritual Guidance for Secular Affairs

The separation of church and state meant that the affairs of state were to lie outside the proper sphere of religion. It also meant that no spiritual guidance was sought on any public policy and administrative issue. Further, somehow it was assumed that while religion dealt with the world hereafter, politics and administration were mostly to concern life in this world. This separation of church and state (which was not universally shared by all world religions and cultures) finally influenced the twin domain of politics and administration when the dichotomy between the two became an accepted fact in the teaching, training, research and practice of the two professions.

Of course, the rise of scientificism, with emphasis on rational objectivity and quantification, equally affected the schism. Slowly, amoralism started to influence the domain of politics and administration, and their decision-making system. The process which began by hailing the separation of state and church went perhaps too far in the other direction, so that the neglect of ethics and values in the education, research and practice of statecraft was either condoned or at best never questioned. It was not realised that the absence of spiritual guidance would accelerate the process of amoralism, selfishness, individualism and materialism. That process, when applied to the secular domain, gives birth to ethical relativism, and removes or weakens such desirable attributes of public morality as individual self-discipline, sacrifice, compassion, justice, equity and striving for the highest good. These attributes draw on the spiritual guidance which world religions and cultures provide. Thus, such a guidance is a necessary and desirable condition for a moral government.

> If justice, equality, equity and freedom are to be maintained, proximate political and administrative acts must have moral standards by which they can be judged. All government acts, if they are to serve the present and future generations well, must be measured against some higher law. That law cannot be a secular law because it is framed by imperfect people in their limited capacities and therefore limited in vision. That law has to be, perforce, based on the principles of higher spiritual and philosophical foundations.[19]

For too long those foundations have been weakening and are in a state of desuetude. The time has come for spiritual guidance to be respected in the matter of public policy and state affairs.

Combating Growing Public Scepticism about Government Credibility

In the past, corruption in politics and administration did not receive the same kind of public scrutiny. Now even a slight deviation from the perceived official norm becomes a matter of great public concern and heated debate. There is clearly in the 1990s a deeper cynicism in the public about the credibility of government. Consequently, any remote possibility of a conflict of interest is seen by the public as yet another example of deception and unethical activity in government.

Not surprisingly, there is a great deal of scepticism concerning the morality of governmental action and the ethics of its officials, and unless public trust and confidence are restored, moral government will remain a distant dream. That restoration will depend upon how public officials are able to demonstrate the highest standards of ethics and responsibility in the use of the power entrusted to them. Towards this, they (the elected as well as appointed public servants) will need to reinforce their sense of mission to serve the public, by acknowledging the place of administrative theology, and by acquiring the necessary education and training in public service ethics and responsibility. Such is the task ahead for those who wish to be involved in the complex and overly-sensitive world of statecraft.

Towards a Moral Administration

Moral administration does not mean that its officials exhibit only the negative obligations such as to do no harm, to avoid injury or to keep out of trouble. On the contrary, the notion of public sector ethics suggests that administrators actively undertake actions that are socially just and moral. Only by actively pursuing the goals of social justice, equity and human dignity can the officials and the state be moral and just. In the past, there was emphasis on negative obligations and admonitions that warned public officials to avoid waste, injustice and the abuse of authority. Today, in the mid-1990s, the administrators also need to subscribe to ethical ideals and personally exercise moral judgment in their duties.[20] By demonstrating the highest standards of personal integrity, honesty, fairness and justice, which are the key ingredients of a moral government, public officials can inspire public confidence and trust.

These ingredients can be strengthened by at least three means. Education and training in virtue and morality are prerequisites to holding a public office. The second means to strengthen the moral resolve of a public servant is the existence of a code of conduct and ethical guidelines to assist a public servant in resolving a possible situation of ethical dilemma. Public servants should be aware of the conduct expected of them, not only by the state which employs them, but also by the public. A set of principles and guidelines with the force of law should be developed, and an office be established for enforcement and resolution of conflicts. Finally, there is a need to resurrect the concept of service and vocation so that a public administrator can rise above individualistic leanings and become a true 'public servant'. Exemplary action against improprieties and malfeasance committed by either the elected or appointed officials would have a salutary effect on the morale of all concerned, including the public.

It should be noted that the morality which determines political and administrative action is multi-dimensional. It is rooted in our civilisation, and derives from its spiritual foundations. It draws from the community of nations and various cultures, and influences the universe that we know of:

> Confidence and trust in a democracy can be safeguarded only when the governing processes exhibit a higher moral tone, deriving from the breadth of morality. This calls for a commitment on the part of elected and appointed officials to moral government and administration.[21]

An exemplary public servant is not simply one who obeys the laws and behaves within the confines of legality, but is also one who strives for a moral government. Such is the duty for those who are members of the difficult and complex world of government. This is the essence and basis of a moral state.

Notes

1.   James S. Bowman, 'The Management of Ethics', *Public Personnel Management*, Vol. 10, April 1981, p. 61.

2. For further discussion, see O.P. Dwivedi, 'Development Administration: Its Heritage, Culture and Challenges', *Canadian Public Administration*, Vol. 33, No. 1, Spring 1990, pp. 94–6.

3. O.P. Dwivedi, 'Bureaucratic Corruption in Developing Countries', *Asian Survey*, Vol. 7, 1967, p. 248.

4. This section and the definition has been drawn from the introductory chapter by O.P. Dwivedi and Joseph G. Jabbra in their edited book, *Public Service Accountability: A Comparative Perspective* (West Hartford, Conn.: Kumarian Press, 1988) pp. 5–8.

5. On the bureaucratic accountability system, see Barbara S. Romzek and Melvin J. Dubnick, 'Accountability in the Public Sector : Lessons from the Challenger Tragedy', *Public Administration Review*, Vol. 47, No. 3, May/June 1987, pp. 228–9.

6. O.P. Dwivedi 'Moral Dimensions of Statecraft : A Plea for an Administrative Theology', *Canadian Journal of Political Science*, Vol. 20, No. 4, 1987, p. 709.

7. O. P. Dwivedi, 'On Holding Public Servants Accountable', in O.P. Dwivedi (ed.), *The Administrative State in Canada* (Toronto: University of Toronto, 1982) p. 173.

8. Joseph G. Jabbra and Nancy W. Jabbra, 'Public Service Ethics in the Third World : A Comparative Perspective', in W.D.K. Kernaghan and O.P. Dwivedi (eds), *Ethics in the Public Service* (Brussels: International Institute of Administrative Sciences, 1983) p. 40.

9. Concerning the impact of the administrative state, see Emmette S. Redford, *Democracy in the Administrative State* (New York: Oxford, 1969); Dwight Waldo, *The Administrative State* (New York: Ronald Press, 1948); and Dwivedi (ed.), *The Administrative State in Canada.*

10. See Gerald E. Caiden, 'Ethics in the Public Service: Codification Misses the Real Target', *Public Personnel Management*, Vol. 10, 1981, pp. 146–52.

11. See O.P. Dwivedi, 'Ethics and Administrative Responsibility', *Indian Journal of Public Administration*, Vol. 29, 1983, pp. 504–17.

12. This discussion has been drawn from author's 'Moral Dimensions of Statecraft: A Plea for an Administrative Theology', *Canadian Journal of Political Science*, Vol. 20, No. 4, December 1987, pp. 703–7.

13. Dennis F. Thompson, 'The Possibility of Administrative Ethics', *Public Personnel Review*, Vol. 45, 1985, p. 555.

14. Dwivedi, 'Moral Dimensions of Statecraft', p. 703.

15. *Bhagavad Gita*, chapter 3, verse 20.

16. Hannah Arendt has demonstrated the absence of any moral qualm among those public servants of the Third Reich who knew about the transportation and murder of Jews. See *On Violence* (New York: Harcourt, Brace and Jovanovich Publishers, 1969).

17. Dwivedi, 'Moral Dimensions of Statecraft', p. 707.

18. Ibid., p. 708.

19. Ibid., p. 705.

20. For a discussion of ethical ideals of public service, see Rosamund M. Thomas, *The British Philosophy of Administration* (London: Longmans,

1978); and Andrew Dunsire, 'Bureaucratic Morality in the United Kingdom', *International Political Science Review*, Vol. 9, July 1988, pp. 179–91.

21. See O.P. Dwivedi, 'Conclusion: A Comparative Analysis of Ethics, Public Policy, and the Public Service', in James S. Bowman and Frederick A. Elliston (eds), *Ethics, Government and Public Policy: A Reference Guide* (New York: Greenwood Press, 1988) p. 318.

4 Science, Technology Transfer and Underdevelopment

THE TECHNOLOGICAL IMPERATIVE: AN INTRODUCTION

Technology can be seen as a set of interwoven contextual circumstances which condition its generation, application and reproduction. Especially important here is the linkage between the so-called hard or gadget technologies and the soft or intellectual social technologies upon which the former rest. However, there is a close connection between science and technology on the one hand and development on the other, which has been at the very core of both spontaneous and induced, endogenous and exogenous, streams of development thought.

From the perspective of development theory, underdeveloped societies are poor and thus unable to generate capital for investment. They also lack the scientific and technical capabilities to extricate themselves from their own poverty and provide the basis for sustainable development. The idea of scientifically driven development has been a central theme in the West since the Age of Enlightenment. But it is during the post-Second-World-War period that this manifestation has been put to full use, centring on the concepts of the diffusion of capital, ideas and technology as the key to modernisation. Here, science and 'manifest destiny' became interchangeable. The roots of this vision go back to the transformation of the world system in the sixteenth and seventeenth centuries,[1] when western technology supported military, economic and ideological ascendancy to conquer the traditional societies of Asia, Africa and the Americas. In the process, the intimate relationship continued between technological innovation and development in general.

TECHNOLOGY: ITS MEANING, NATURE, AND SIGNIFICANCE

Technology can be said to be the use of scientific principles, together with human skills and man-made tools, to manipulate material objects and physical forces in order to accomplish set goals. Denis Goulet has defined this further:

> Technology may be defined as the systematic application of collective human rationality to the solution of problems by asserting control over nature and human processes of all kinds. Technology is normally the fruits of systematic research which is disciplined and cumulative, not merely accidental and serendipitous. Moreover, it is not mere intellectual speculation or theoretical modelling but rather knowledge *applicable* to practical problems. And further, this systematically applied human reason must operate in a *collective* societal context, so that a practical invention which originates in a solitary mind does not qualify as technology unless it is expressed in a tool, process or object which can be used by others. Technological activity, then, aims at expanding and improving the ability of human beings *to control* the natural and social forces which surround them.[2]

There is a close link even between science and traditional technology. The nature of such a link is illustrated from the following example drawn from medical practice in Africa. A traditional medicine man knows, from having seen it several times before as practised by his family elders, that a potion prepared from a particular herb will cure a particular disease. But he knows nothing of the nature of the chemical action of the potion on the malady. He has not seen any need to analyse the constituent elements of the herb in question. He does know, though, the external manifestation of the disease. He is completely ignorant but is able to inspire the necessary confidence in his patients who know that they are cured. All he knows is that, whenever he perceives those symptoms and administers this potion, the sufferer's health is generally improved.[3]

In this case, the medicine man knows 'how' without knowing 'why'. This is a good example of the use of technology without a deep background of scientific knowledge and laboratory facilities. The following illustration, also drawn from an African setting, substantiates the above point. Suppose an African student doctor, researching for his MD, carried out an investigation into the healing herb of his less enlightened brother, and analysed its chemical elements; suppose he further carried out experiments and found out that the disease was not merely functional but organic, and discovered the microbes that caused the malady and how they affected the organ they attacked; and suppose he carried his experiments further and determined precisely which of the chemical constituents of the herb was death to the germs and health to the patient. He would then know both the nature of the herb, the nature of the disease and the precise reason why the former cured the latter. 'Thus he knows not merely that a certain herb cured a certain disease, but also how it did so. In brief, he knows not only the phenomenon, but also its cause.'[4]

This is, then, the nature of the link between technology and science. Science provides the causal principles that are the creative foundations or springboard of knowledge-based invention; on the other hand, an existing invention challenges man's innovative genius. Realising this connection, governments of several countries have established institutions (such as for Ayurvedic medicine in India, Unani and Tibbiya medicine in some Islamic nations) to conduct research into traditional systems of medicine and health care so as to turn that folklore medicinal technology into a modern medical cure.

Technological Gap between North and South

G.E. Skorov has pointed out the historical forces and processes which combined to undermine Africa's technological development:

> There was a time when the peoples of a number of Asian, African and Latin American countries held leading positions in the world of science and its practical application. Major discoveries in astronomy, mathematics, and medicine, and achievements in the art of building, in irrigation and sea-faring stirred the early Middle Ages in China, India, Egypt, South-Western Asia and Central America. But, because of certain specific features of the historical process, the peoples living in these regions experienced a slower rate of advance in the Middle Ages, while during the colonialist period, their independent development was interrupted. The creative drive of these enslaved peoples was fettered and their cultural life paralysed. The colonies and dependent countries were not only unable to participate in the development

of world culture, science and technology but they even lost much of what they had themselves created while they were still independent.[5]

This, then, is the origin of the technological gap between the South and the North, the negative consequences of the technological disparity on the South being most pronounced in the economic domain. There are three main reasons for this: first, technology plays a cardinal role in the transformation of material resources. In the absence of such a technological capacity, the Third World countries have depended on the multinational corporations to exploit their natural resources. Secondly, technology is an 'appropriated science'. It has a price tag and can be bought, sold or loaned, like other commercial items; but most importantly, it is a protected value with patents, trade marks and related constraints. Thus it is not easily and cheaply available to the scientists and professionals of the Third World. That is why the supply has always come from the North and the terms and conditions have generally been harsh on the South. Thus, the North–South technological disequilibrium has prevented the less-developed countries (LDCs) from making effective use of their natural resources. It has also made it impossible for these countries to predicate the development of their economies on local factor endowments, with the primary objective of catering to society-wide basic needs.[6]

The third reason relates to the pressures being brought upon the South by the industrialised nations to keep their natural resources intact in the name of saving the planet from global warming and further depletion of the ozone layer. Yet when it came to the protection of our planetary biodiversity resources, countries like the United States at first refused to sign the treaty drawn up at the June 1992 Earth Summit. It seems that the South is being asked to pay for the environmental excesses committed by the North when the latter were able to industrialise and reached a quality of life which is the envy of the South.

THE CONTEXT OF SCIENCE AND TECHNOLOGY

A scientific and technological system is not a closed system. Paraphrasing Ortega y Gasset's famous dictum, such a system could be seen as itself, and its circumstances.[7] Not only are science and technology increasingly interconnected, but information about problem-solving (or know-how) relates to the circumstances where such know-how – as well as know-what and know-why – is generated. This means that scientific and technological systems are immersed in and part of a complex social, economic and political world which gives technologies and techniques social meaning and orientation.

Several concepts are central in this analysis. The first is the notion of technological *product*. This is a material tool or gadget (for instance, a piece of machinery or equipment) which is used to deal with a problem. This tool is a means to approach an end (the solution of the problem). The relation between ends and means is not always a simple one, especially under conditions of technological dependence, and also because similar tools may be used to tackle diverse problems. Secondly, the tool is a product of a technological *process* – the creation and innovation of tools – which is inserted in a body of knowledge or experience, whether folk or scientific. A technological product is not the technology; for the latter to exist there must be a capacity of production, reproduction and innovation. The third notion is that of a technological *function*. We mean by that a combination of tools, operators (those who have learned a technique), a politico-administrative infrastructure (including patrimonial, bureaucratic and participatory practices), and culture (including beliefs, folk knowledge and science), all oriented to solving a problem. A series of technological functions to tackle a problem and their interconnections makes a technological *matrix* or a combination of functions. Finally, there is the notion of technology as a juxtaposition and articulation of technological matrixes which has a built-in capacity for expansion and reproduction of know-how.

Thus, when we talk about a society's technology, we are referring to a complex and dynamic pattern of interconnected technological functions and matrixes. These represent in broader terms the society's 'ways of doing things', or problem-solving. Technology, therefore, extends from the tools, instruments and machines to those who possess the skills or techniques for their use. It encompasses also the broader social organisations in which operators apply their skills as well as, most importantly, the cultural-educational institutions for the creation, training and reproduction of know-how.

A technology is a system for problem-solving where five major elements are closely interrelated: a material and perceived context or problem which needs to be addressed; a culture which gives meaning –

purposes, feelings, cognitions and expressions – to such a context; a structure of groups and individuals with resources and networks of communications charged with dealing with the problems; a set of processes whereby groups and individuals act in the pursuit of purposes; and the effects or consequences of these actions, in this case the overall environmental impact related to development. Needless to say, these are only some of the multifaceted and complex factors at play in the technological process. For instance, the distinction between 'national' and 'international' is rather artificial. In practice, there is a multidimensional continuum ranging from domestic to national, from the informal to the highly institutionalised, from grass roots to high-level. This simplification, however, may help to suggest possible and often contradictory relationships in technology generation, application, its socio-political effects and feedbacks.

Factors Influencing Technological Systems

For the purpose of simplifying the relationships between scientific and technological systems and their environments, we have made an equally arbitrary distinction between cultural, professional and politico-administrative environments. Needless to say, these three contexts affect not only science and technology but are also closely interrelated among themselves. Perhaps one could represent these ambiences or circumstances as a series of concentric circles, at the outer rims of which there are the problems which are being dealt with by technological tools. Underpinning such technological tools, however, there are those social groups specifically trained to operate and apply those technologies. These groups are the professionals or, in a broader sense, the technocracy. Such a technocracy, especially since the managerial revolution, does not operate in isolation. It is supported by an administrative (as well as political) system of organisation and management for the delivery of its technologies (for example, in the medical, educational, architectural, engineering, military, and agricultural sectors). We could refer to this layer as social technology. Finally, there are the complex cultural foundations (values, beliefs and symbolic representations) which give meaning to the social arrangements and gadgets of a society. I will examine these interrelationships in some detail below.

*Cultural Circumstances of Science and Technology in Third World
Countries*

From a broad perspective, a technology can be seen as part and
product of a culture.[8] The idea of culture encompasses a burden of collective experiences – ways of going about doing things, artifacts, folklore and other spiritual and/or artistic expressions. It comprises the
complex gamut of a society's reconstruction of reality, from systematic knowledge (philosophy and science) to feelings, morals and intui-
tions. In an Aristotelian way, we could say that a culture includes the
collective orientations of all three aspects of the human mind: know-
how, the creative knowledge of aesthetics, intuition and inspiration
(art and literature), and the world of abstract and systematic thought
(theory and science).

The visible expressions of a culture are material: artifacts, utensils,
buildings, works of art. Its foundation, however, is language and symbols which could be combined, recombined and modified. A culture
involves knowledge broadly defined: know-how (technology), know-
what, know-why and know-when. The latter three in modern society are
synonymous with science. It is perfectly possible, as has been the case
in human history, that know-how could develop independently from science.[9] In fact not only in the history of civilisation has technology preceded science, but it has often provided the problematic basis for
scientific discovery. However, at present the capacity to deal with complex problems and to retool such know-how requires a systematic
understanding of reality. In this sense, as Comte suggested, we can talk
about a scientific, as opposed to a theological or metaphysical, culture.
The leading edge of technology nowadays is given by research and
development (R&D). This is a relatively novel phenomenon which is
particularly pronounced in the most advanced industrialised societies.[10]

The cultural circumstances of science and technology in Third World
societies are different from those prevailing in the industrialised and
post-industrialized centres. In the first place, science as a cultural component is often associated with First World indigenous culture. It may
well be perceived as a form of cultural imperialism.[11] In fact, nowhere
is this false dichotomy of modernity and tradition more manifest than in
the cultural sphere. Science in this case could become either a cultural
component to be rejected by traditionalists and radicals alike, or an
intellectual fetish to be used by westernised elites in order to legitimate

their superiority. The phenomenon we are discussing here is that of cultural alienation, and more specifically the alienation of science from Third World culture. As a consequence, scientific and technological systems in many Third World societies will be perceived as – and in fact they may well be – divorced from their own culture as well as from the set of concrete and pressing problems which need to be addressed.

This is not to say that Third World societies do not possess scientific and technological systems of their own. However, their own cultural legacies and experiences are often neglected and pushed aside as irrelevant, primitive or unfashionable. A final point regarding the phenomenon of cultural alienation is in order here: underdeveloped societies may be poor societies but not necessarily under-intelligent societies. Most often, cultural transfers and the unrestricted acceptance of modernity reduces more than enhances the possibilities of scientific and technological innovations. This results in a pattern characterised by the following traits:

(1) There is a lack of autonomous scientific and technological capacity. The initiative for innovation, R&D, and so on is left in foreign hands.
(2) Most science and technology is a commodity owned by foreign countries.
(3) There is a disjointed or interrupted connection between scientific research and technological development.
(4) There is a lack of correspondence between science and technology and their circumstances.
(5) There is a discrepancy between indigenous and traditional social technology and imported technological products.
(6) Last, but not least, there is a growing gap between the problems to be solved and the often imported and a priori technological solutions.

Imported solutions do not create an autonomous technological capacity *per se* but they do alter the existing social technologies in an unintended, unplanned and uncontrolled way. It is also important to re-emphasise that using a gadget does not imply possessing the technology.[12] A society can become used to technology but not necessarily be technologically minded, in the sense of producing its own answers to its very own problems. R&D is fundamental here and this means theoretical research. Theory and praxis are dialectically interrelated and this is particularly true since the Second World War. Third World countries cannot break the technological trap until this fundamental relationship is recognized and acted upon.

Professional Circumstances of Science and Technology in Third World Countries

The operation and application of a scientific and technological system (and for that matter of any individual technology) is the realm of professionals and scientists which correspond to two different levels of post-secondary education and training. Before the advent of modern capitalism, not only were science and technology quite separate in the western world, but technology was the domain of a complex web of guilds and corporations trained by apprenticeship and practice. The modern system of tertiary education is chiefly a creation of the French Revolution,[13] as exemplified by the Napoleonic university and the Polytechnique, which blended scientific and technological training. Modern professions emerged in the nineteenth century and spread by imitation throughout Europe, North America, Japan and to what was later to be known as the Third World. With the advent of the administrative state, what used to be called the liberal professions increasingly encompassed a corporate technocracy charged with the solution of complex social problems. Professionals, as operators of technology (and technique), tend to exist nowadays immersed in large-scale organisations, whether private or public. It is the institutions of higher learning which have a primary responsibility for the creation and reproduction of such a technocracy.

The above is true for both developed and underdeveloped societies, although in the case of the latter, the technocracy is affected by other characteristics. For example, in the absence of an autonomous scientific and technological capacity, training tends to be imitative. Further, there is a transnationalisation (meaning westernisation) of professional standards, and there is a severe incongruity between professional training and scientific research. Also, the problems (such as research on bio-seeds, cancer, and diseases) addressed by science and technology tend by and large to be over-determined by the importation of know-how. That means that the circumstances of science and technology tend to be imported, and surreptitiously. Finally, there is a profound discontinuity between what is often referred to as traditional ways of problem-solving and the modern imported methods. The most fundamental issue, however, is the above-mentioned lack of an autonomous and self-reliant scientific capacity. Third World professionals tend to be well-trained, often over-trained, but unable to operate in their own environments. They contribute in no small manner to the advancement of the scientific and technological capacity of the developed nations through the process known as the brain drain. Needless to say, this has the most devastating effects for the expansion, even survival, of scientific and technological systems in Third World societies.

Politico-administrative Circumstances and Technological Underdevelopment

While the cultural and professional circumstances of scientific and technological systems in the Third World are important, their administrative and political circumstances need to be especially highlighted. Three major reasons are suggested here:

(1) The first is that not much is known about the connection between science and technology on the one hand, and the practice of government on the other. Although science and technology require a great deal of administrative capability, this is largely assumed to be a dependent and circumstantial factor, subordinated to gadget innovation.

(2) The second reason is that politics and administration are essential components of a technological function, and certainly of any technological matrix. In fact, without a political framework (including purposes, alliances, institutions and policies) and administration, we could not talk about a technological function at all. Especially in complex societies, technological products and professional techniques without planning, supports, policies, organising and management would be lacking a will, a delivery system and an infrastructure to handle social problems.

(3) The third and most important reason is that political and administrative know-how – or social know-how – is technology, no matter how implicit or soft. Moreover, it is precisely in this area that developing countries have the capacity and potential to exercise uninhibited innovation. An explicit, rational and appropriate politico-administrative technology – appropriate, that is, to the

needs and conditions of the country – has the capacity to harness foreign technological products and create distinctively national technological functions to address national problems. Development projects and administration are precisely technological functions.

A number of general observations regarding the training and professionalisation of development administrators is in order here. These are not intended to provide a thorough list of administrative and political constraints affecting Third World societies. For one thing, it is extremely difficult to generalise about a set of regions of the world which encompass about two-thirds of humankind. These factors are a crude reminder of two essential analytical assumptions: one is the foundational nature of soft political and administrative technologies for any technological function and/or development plan; the other is the necessity to treat political and administrative categories as inextricably linked social know-how, rather than from an antiseptic perspective separating politics and management.

(1) The most neglected area in the application of politico-administrative technology in Third World societies happens to be the rural areas and at the grass-roots level. Moreover, these are precisely the areas where the development effort is most needed and where the greatest potential for self-reliant social mobilisation exists.[14] Therefore, a strategy oriented to enhancing developmental effectiveness must emphasise, by necessity, the activation of the peasantry.

(2) Social technology transfer from the developed to the underdeveloped world is by and large not appropriate. Despite the acknowledged fact that technical assistance plays an important part in Third World development, more often than not technology transfer from the West has had a less-than-salutary effect on the recipient country. This dysfunction tends to emerge from the common practice of having found the solution before defining the problem; or worse, trying to adapt the problem to a standard solution. Therefore, a transfer in managerial know-how is more effective when a self-reliant capability is created.

(3) Politics, in the sense of the social capability for mobilising, motivating, integrating, for decision-making, as well as for communicating and creating bonds of trust and confidence, is at the core of material and technological development. The creation and

maintenance of credible leadership as well as the empowerment of diverse social actors are not simple engineering-like functions. Issues of public morality, respect, rectitude, ethics, responsibility and accountability are interwoven with questions of effectiveness, efficiency and professionalism.[15] Because the administrative, political and technological cultures are interconnected, efficiency without political meaning is void of content. Conversely, political projects and energies without instrumentalisation in concrete and sustainable results are either rhetoric or failures.

Most Third World societies, however, are not characterised by high levels of legitimacy and effectiveness. If anything, their pattern of disjointed development has created a political gap. The inability to constitute political communities, incorporate popular demands, institutionalise politico-administrative practices, deliver results and maintain sovereignty are endemic to the peripheral state. Present practices tend to reinforce the dependent, underdeveloped and exploitative nature of the existing socio-economic order. By and large, the so-called developing societies are not developing, while their so-called development administrations are contributing in no small measure to the administration of underdevelopment.

A two-tier system of administration, where a formal veneer of public service co-exists with predatory practices, often emerges. This results in entrenched systemic corruption and a decline of legitimacy. The tendency to substitute neutral bureaucrats or technocrats for the more political leaders – a practice rooted in the aforementioned perception of corruption – often fails to produce effective management or even to reduce corruption. The substitution of one ideology ('technocratic') by another ('political') does not change significantly the values of the state classes unless accompanied by a major structural socio-economic change. That is, the administrative culture is a symptom, not a cause, of deeper systemic circumstances. The elimination of politics, far from solving the problem of conflict-management and underdevelopment, entrenches violence, injustice and poverty. From this viewpoint, the two basic conditions for the effective utilisation of technology are the question of who wields state power and, most important, the issue of rapid and equitable political and socio-economic change.

(4) Politico-administrative sovereignty – the capacity of a society to diagnose its own problems and devise its own collective course of

action – is an equally important constraint on technological development. Most Third World countries lack economic and even political sovereignty. Often the administrative classes on the periphery either reflect the values of the former colonial bureaucracies, or are parasitic, or both. Commodity or rentier states are particularly vulnerable to the transnationalisation of sizeable sectors of their military and civilian bureaucracies through trade, foreign aid, training and other bilateral mechanisms.

Strategic considerations of this nature play a major role in the international transfer of technology. Most Third World countries not only receive derivative technologies, but proprietary controls limit the adaptation and generation of know-how. Moreover, especially in the area of high-tech transfer, the predominant gadgetry is military hardware: the most useless, wasteful and dependency-ridden of all technologies.

In sum, the same characteristics extant in the cultural and professional circumstances of science and technology in the Third World are present with regard to its politics and administration. The politico-administrative systems on the periphery tend to portray a series of characteristics which inhibit development. Among these, the imitative, ritualistic, corruption-plagued, un-coordinated and generally deficient nature of the Third World state constitutes the fundamental barrier to technological development. Any analysis of appropriate administrative technology – and of course prescriptions – has to start from an understanding of these politico-administrative limitations.[16]

CONCLUDING REMARKS

In their quest to industrialise fast, to modernise their economies, and to ameliorate poverty, the developing countries, over the years, have been importing technology on a massive scale from the advanced countries. Firmly believing that industrialization will bring material progress as well as a higher standard of living, the developing countries decided to opt for the pattern of development that would replicate the economic, social and political modes of the industrialised countries. Because such progress is not feasible without the use of science and technology, the transfer of technology has become indispensable. This, in turn, has meant acquiring equipment, product and process designs, management techniques, technical skills and scientific knowledge. Thus, technology transfer entails the transplantation of technical knowledge and material, and also the creation of the foundation of the basic sciences. The foundation includes the acquisition of basic scientific knowledge freely available through professional training and research publications, as well as the development of the necessary infrastructure, including a pool of indigenous scientists and engineers who can apply that scientific knowledge to their specific conditions.

However, technology transfer, contrary to popular belief, is not value-free. It has certain inherent cultural biases and distortions which are also transferred. Too often recipient countries have become dependent on a kind of technology that has led to negative, unforeseen consequences. In addition, the technology itself, transferred through the subsidiaries of transnational corporations, has increased Third World dependence on the import of capital goods and intermediate products and components, and the use of foreign services and consultants. This has created technological over-dependence in many of the developing countries.

It should also be realised that as the social, political and cultural natures of the Third World countries differ, each Third World country would have to chart its own plan and strategy for technological progress, by seeking technology and improvising and grafting upon its indigenous capabilities. The strengthening of indigenous scientific capability requires development of the necessary basic theories and assumptions. But the Third World countries and their scholars have not yet produced their own theories, methodologies, and paradigms of development. Without building such an indigenous theoretical base and methodological foundation, any transfer of knowledge and tools from vastly different socio-cultural settings may not always produce the desired results. Only with the development of their own methodologies and assumptions can the transfer of Western science and technology act as a midwife assisting in the natural delivery of desired progress.

One major factor in securing that desired result will be the transfer of managerial know-how when a self-reliant scientific research capability is created. There is an acute shortage of managerial technology in the Third World societies. Training for development management would be required so that these societies are able to create and maintain a human capability for problem-solving, and establish and manage a scientific and technological system precisely geared to generating and reproducing such capabilities.[17] For this, these countries should consider securing help not only from the West but also from among themselves (especially from those nations who have been able to create their own indigenous scientific capability).

At present, there is an urgent need for a greater exchange of information among the Third World countries with respect to each other's technological capabilities and requirements. The Third World nations should build educational and research facilities for scientific and technical knowledge which could be oriented towards their own needs, because the benefits of mutual transfer of technology are indeed very great.[18] Some countries, such as India and China, have made considerable progress over the years, and this experience can be put to good use by other developing countries.

In addition to mutual self-help, intellectual co-operation between North and South is needed so as to develop mechanisms and methodologies for the transfer of intellectual technology, strategies for technological development (both regional and national), desired high technology for the Third World, and more particularly appropriate technology to protect the environment, health and welfare of their people. If we believe that science, like truth, is a common heritage of our global community, its technology should then be available to all, without constraints. Of course, if the West is going to be magnanimous in releasing the technology from its bondage, the South will have to act responsibly in the use of such technology.

POSTSCRIPT: EMERGENT ISSUES

There are several issues which have emerged in the early 1990s and which require our immediate attention during the next few years.

Transfer of Intellectual Technology

In the mid-1970s, a criticism arose that the Western concepts, models, paradigms and theories were inappropriate for understanding the completely different circumstances of developing countries. The suggestion was made that such alien concepts and methodologies resulted in inappropriate policy decisions, and diverted the attention of the leaders from the real problem of fulfilling the basic needs of their people. Heavy emphasis on GNP per capita, capital-intensive projects, value-free and culturally neutral know-how, and consumerism actually misled the policy-makers of the Third World into adopting a planning mode which had no place for indigenous capabilities and resources. Thus, the paradigms of the West served as blinkers or even as escape mechanisms preventing the policy-makers from realising and implementing appropriate strategies. The domination of the West in the field of intellectual knowledge and methodologies continued the dependence of the Third World. Exceptions to this were countries such as India and China. The impression that the developing nations could ill afford investment in basic sciences strengthened the concept of centre and periphery. It also stunted the hope for achieving self-reliance, self-confidence and autonomy in the field of science and technology.

So far, western intellectual domination remains a fact of life. What is most surprising is that the Third World scholars who have been trained in the West still continue to seek recognition and funds from the industrialised world. The only duty of these scholars (including those who have settled nicely in the West) is to develop alternative models and strategies deriving from their cultures and background, and relating to their specific needs and resources. If there has been a bias in transferring intellectual technology from the West (and why should that bias not exist?), the subservient attitudes of some Third World intellectuals is also a contributing factor. One way to balance the situation is to recognise that all scientific and technological discoveries and innovations are universal truths, but that none of these truths are free from their inherent cultural underpinnings. Just as forms of worship differ among world religions although the ultimate goal of individuals may be the same, so too methodologies of achieving development differ although the goal may be the same. It is imperative then that, while transferring technology, both the concepts and tools be transferred; it is up to the recipient to decide what form or methodology is to be applied. Unfortunately, this is too often not the case.[19]

It is of the utmost importance that researchers of the West working in developing countries include a scholar from the Third World as an equal partner in their research. The intellectual capabilities of indigenous people is of an equal standard, and that should be the basis of collaborative research. Of course, such collaboration becomes much more honest and appropriate if both partners gain something from it. Complementarity is necessary in securing participation and collaboration between researchers of the West and the Third World. In this way, one can create an appropriate balance in the transfer of technology, both intellectual and otherwise.

The Issue of Bio-diversity

In Rio de Janeiro in 1992, the United States refused to sign the bio-diversity convention, though the Clinton administration reversed this decision. The developing world is a repository for the world's genetic pool. Of the eighty million species surviving in the world, 70 per cent are said to be found in the tropical zone. In the case of India, it claims to have 45 000 plant species including 15 000 species of flowering plants, and 75 000 species of animals including 50 000 insects, 2000 fishes, 420 reptiles, 1200 birds, and 340 mammals and other invertebrates.[20] The convention asserts that 'recognizing the sovereign rights of states over their natural resources, the authority to determine access to genetic resources rests with the national governments and is subject to national legislation' (article 15-1). The United States felt that this clause would deny its multinational corporations access to the vast gene pool of the Third World. Such a binding commitment, it is feared by the United States and its allies, could be used by the Third World to put pressure on the industrialised nations to revise GATT and other agreements on patents and intellectual property rights. If the bio-diversity treaty is accepted by all, then many of the poorer nations would benefit immensely. For example, Peru could receive millions in royalties due to it for the potato gene it gave to others. Similarly, there are other developing nations which may receive similar royalties on their genes which were taken away by the West in the past.

Overdevelopment of the North

The economic system of the North is geared towards generating non-essential goods and services in a manner needed to maintain a very high-consumption lifestyle. With a culture of built-in obsolescence, a superfluous and wasteful consumption pattern is sustained, and nowhere is this wasteful style more evident than in the North's agricultural policies, which are ecologically destructive and energy inefficient.[21]

This overdevelopment in the North is accompanied by underdevelopment elsewhere. While a growing majority of people (mostly in the South but also increasingly in the old command economies of the Eastern European nations) are unable to satisfy their basic human needs, a major portion of resources is being exclusively used up by the small proportion of the world's population in the North (and also by a tiny group in the South). Production processes and the emphasis on monoculture farming, which deplete world resources and destroy biodiversity, ought to be changed. As a matter of fact, the primary responsibility for global environmental security and peace rests with the North. That is why it becomes their primary responsibility to reduce global ecological stress by dealing with their unsustainable consumption patterns.

Scientific and Technological Dependence of the South

Northern technology is capital- and energy-intensive with a very low labour input. And this is what is easily available and marketed by the North. The Third World countries are often propelled into new technological territory which their own indigenous framework, management system and scientific base cannot effectively sustain. The southern nations do not realise that the technological package which is being transferred to them was developed, in the first place, for specific climatic and consumption patterns in the North, where conditions are different.

Success in the North does not mean that the same technology will work equally well elsewhere. For example, the 'Green Revolution' package was hailed at the time as a miracle to solve all the food problems of the South. Since its introduction, it has helped to destroy the gene-pool of indigenous varieties of wheat and rice, promoted an excessive use of chemicals, depleted soil nutrients, induced pest immunity, and generally created ecological degradation. Another example is the displacement of the traditional method of fishing by the use of modern trawlers and equipment which were provided to some developing nations as a part of the aid package, and which have now created havoc with the fish stock within their territorial and economic zone waters. Finally, there is another type of technological dependence which is not being discussed openly: this relates to the 'export' of scientific data from the South for analysis and use in the North. In some cases, the research by non-nationals remains unregulated, so that the research methodologies and results are frequently not shared with the local researchers of the South. The problem of inequitable participation in research and development ought to be rectified soon.

Planet Protection Fund

During the Earth Summit, a proposal to establish a planet protection fund was advocated by India.[22] Such a fund could be used to restore the environmental health of the globe and save it from virtual environmental doom. The plea is to make all nations share an equitable financial burden for assessing environmentally friendly technologies. The proposal envisages that states would contribute at least 0.1 per cent of their GDP for the fund. Only the least developed nations would not pay into the fund, which would be managed by an international agency such as the UN or a similar body set up for the purpose. It would ensure that research and technology development pertaining to environmentally friendly technologies would be made available to any nation without the constraints of patent or trade restrictions.

Technology: a Common Heritage of Humanity

There are some basic questions which we ought to ask. Should not technology (akin to scientific discoveries) be made a common heritage of all? Should not those international obstacles which impede access for the South to the advantages of modern scientific and technological breakthroughs be removed? These and related questions have been on international public agendas for decades but without any resolution.

Reluctance on the part of the North is based on several factors. First, the major owners of technology – Western multinational corporations – fund an estimated 75 per cent of research and development projects. These corporations naturally expect rewards from their investments, and any suggestion that the fruits of their research should become an item in free gift packages rapidly degenerates into a dialogue in which neither party is willing to listen. Secondly, the North would lose the advantage of value-added profit and this could appreciably reduce the profit margin of its products on the international market. A third reason for northern reluctance is that an industrial South could limit or even close access to the sources of raw materials for northern industries, thereby shrinking the southern market for products and obsolete technology of the North.

Nevertheless, a technologically weak and economically poor South will remain a burden for the North. That is why it is in its own interest that the North takes steps to remove the ghettoisation of the South; otherwise consequences could be disastrous for the entire planet. Additionally, a technologically and industrially stronger South is likely to increase the earnings of southern consumers and thereby enhance their power to purchase northern goods. Also, such a change is likely to reduce the technological gap, create a more equitable economic order, and reduce North–South tension. Finally, there is a need for a global restructuring of the existing technological order to enable the South to use its resources more efficiently and thereby partially resolve such world problems as food and energy shortages, and environmental pollution.

Gandhian Philosophy versus Western Technology

Perhaps the only modern ideology that rejects the western style of consumption patterns and materialistic values, and which provides us with an alternative model to fill the conceptual vacuum in the South, is Gandhism. Gandhian philosophy is based upon a non-materialistic, anti-poverty and non-industrial economic system; the transfer of ideas and concepts on a non-ethnocentrism basis; an emphasis on non-exploitation and ecologically regenerative systems (for example, relying on the agro-ecosystem to create the potential for sustainable livelihoods); a decentralized political system and a participatory citizenry which takes initiatives to resolve its needs mostly with locally available resources; and a non-violent social system. It also calls for an emphasis on culturally and ecologically respectful solutions, which do not necessarily decry all western ideas as being totally unsuitable. Instead, it contends that the transfer of technology ought to be based on a proper balance between the traditional knowledge and culture, and the modern technological system being imported from outside. It maintains that the best methodology is one which sets up the most appropriate technology for the attainment of egalitarianism based on material minimums and indigenous self-sufficiency. Gandhian technology is built upon local socio-economic resources, agriculture, cottage industries, labour-intensive productivity and economic self-sufficiency.[23]

Similar views were also advocated by the Commission on Developing Nations and Global Change (p. 71):

The need today is to integrate the ecological wisdom of traditional practices and technologies with the production potentials made possible through modern science and technology, so that increased productivity can be obtained in an ecologically benign manner.

The historic decolonisation process began after 1945, starting with the independence of British India. Though India and most other Third World countries started rejecting western models of development, the question to be asked now is: Have these countries been faithful to their original intentions? Have their latest economic policies and programmes remained consistent with their original ideological pronouncements? They seem to have accepted blindly and totally the western ideology of unlimited materialism, industrialism and militarism. The western experts seem to have convinced leaders of the Third World that progress is unilinear and that the western models of development are, in some sense, the products of evolutionary progress.[24] The ruling elites in the developing countries either do not realise, or refuse to reveal to their peoples, that they can never become materially rich like the developed western states, nor is it desirable to be so materialistic in values, a fact that is now being recognized in the industrialised West itself. The most fundamental reason for the deviation of the Third World from its own ideologies lies in the fact that its leaders falsely believed that western technologies could help them attain their own ideological goals.

As a result, Third World leaders have resorted to a wholesale transfer of technology from the West. According to a UNESCO report published in 1976, 'At the present time the transfer of technology to the developing countries is unprecedented in its scope, rapidity and urgency, and implies such radical changes that some countries regard it as... a form of cultural aggression'.[25] Most of this transfer of technology has been through the instrumentality of so-called foreign aid. An important condition and component of foreign aid has been military aid. In the words of the 1980 Report of the Brandt Commission:

Some Third World countries have substantially boosted their armaments, ... encouraged by arms producing countries.... It is a terrible irony that the most dynamic and rapid transfer of highly sophisticated equipment and technology from the rich to the poor countries has been in the machinery of death.[26]

Development Administration

Western technologies have evolved in different historical, cultural, economic and political circumstances, which do not prevail in the South. We now see the disastrous results of the use of these technologies by the West, and their threat to our planet.

The Third World countries have not yet realised that by accepting, inviting or copying western technology, they are importing western ideologies as well, even while they deny or remain unconscious of such importation. They are unaware of the impact of science and technology upon their native philosophies, religions and cultures.[27] The historical evidence is that similarities of technology have been eliminating even the secondary differences between the western ideologies of capitalism, socialism, communism and fascism. The West cannot and should not be expected to be so unselfish and generous as to design, create and cater to the special needs, economies and ideologies of the Third World, irrespective of the cost involved. Therefore, for the West to design and supply technologies appropriate to the Third World may be seen as unrealistic, inappropriate and unhelpful.[28] What the Third World needs is not an appropriate technology supplied by the West, but an alternative technology it has itself created on the principles of equal partnership, self-reliance, self-sufficiency and self-management.

The environment–development crisis facing our planet has given us a great opportunity to restore and strengthen international cooperation in the matter of science and technology. It is an opportunity to bring together both the will to formulate and implement broad-based and mutually beneficial strategies and the institutional arrangements to ensure not only humanity's survival but also its ability to live in peace and dignity. A new process of adjustment is needed, not driven by market mechanisms alone but by ecological imperatives. There is an urgent need to bring into play a spirit of genuine international co-operation towards environmentally sound and sustainable development. These and related aspects are examined in the next chapter.

Notes

1. See Immanuel Wallerstein, 'Crises: The World Economy, The Movements, and the Ideologies', in Albert Bergesen (ed.), *Crises in the World System* (Beverly Hills: Sage, 1983) pp. 21–36.
2. Denis Goulet, *The Uncertain Promise: Value Conflicts in Technology Transfer* (New York: IDOC/North America, 1977) p. 4.

3. See, for example, the essay by Collins E.N. Ngwa, 'The Technological Variable in North–South Relations: An African Perspective', in O.P. Dwivedi (ed.), *Perspectives on Technology and Development* (New Delhi: Gitanjali Publishing, 1987) p. 122.

4. Bernard Fonlon, *To Every African Freshman: the Nature, End and Purpose of University Studies* (Victoria: Camroon Times Press, 1969) p. 25.

5. G.E. Skorov (ed.), *Science, Technology and Economic Growth in Developing Countries* (New York: Maxwell House, 1979) p. 7.

6. Ngwa, 'Technological Variable', p. 127.

7. This section is based on O.P. Dwivedi, J. Nef and J. Vanderkop, 'Science, Technology and Underdevelopment: A Contextual Approach', *Canadian Journal of Development Studies*, Vol. 11, No. 2, 1990, pp. 227–36.

8. See Richard Israel Zipper, *Un Mundo Cercano. El Impacto Politico y Economico de las Nuevas Technologias* (Santiago: Instituto de Ciencia Politica, Universidad de Chile, 1984) pp. 11–30.

9. Bernard J. Stern, 'Some Aspects of Historical Materialism', in *Historical Sociology: The Selected Papers of Bernard J. Stern* (New York: Citadel Press, 1959) pp. 3–14.

10. See Zipper, *Un Mundo Cercano*, pp. 13–19.

11. See O.P. Dwivedi and J. Nef, 'Crises and Continuities in Development Theory and Administration: First and Third World Perspectives', *Public Administration and Development*, Vol. 2, 1982, pp. 62–3.

12. Nora Cebotarev, 'The Diffusion of Technology, Blessing or Curse for Latin American Nations?' in J. Nef (ed.), *Canada and the Latin American Challenge* (Guelph, Ontario: Cooperative Programme on Latin American and Caribbean Studies, 1978) pp. 97–103.

13. Stern, 'Historical Materialism', pp. 36–44.

14. See A.M. Walsh, 'Public Administration and Development', *Institute of Public Administration Report 2*, No. 1 (New York: Institute of Public Administration, 1984) p. 10.

15. See O.P. Dwivedi and R.B. Jain, *India's Administrative State* (New Delhi: Gitanjali Publishing House, 1985). Also, on public service accountability, see Joseph G. Jabbra and O.P. Dwivedi (eds), *Public Service Accountability* (Hartford, Conn.: Kumarian Press, 1988).

16. For a typology of administrative problems, see chapters by Ferrel Heady, and Gerald and Naomi Caiden, in O.P. Dwivedi and Keith M. Henderson (eds), *Public Administration in World Perspective* (Ames: Iowa State University Press, 1990) pp. 3–8 and pp. 363–99.

17. For further elaboration, see J. Nef and O.P. Dwivedi, 'Training for development management: reflections on social know-how as a scientific and technological system', *Public Administration and Development*, Vol. 5, No. 3, 1985, pp. 235–45.

18. For further elucidation, see T.N. Chaturvedi, *Transfer of Technology Among Developing Countries* (New Delhi: Gitanjali Publishing House, 1982).

19. Paul P. Streeten, 'Problems in the Use and Transfer of an Intellectual Technology', in P.J. Lavakare *et al.* (eds), *Scientific Cooperation for Development* (New Delhi: Vikas Publisher, 1980) p. 60.

20.   India, Ministry of Environment and Forests, *Traditions, Concerns and Efforts in India* (National Report submitted to the United Nations Conference on Environment and Development, Rio de Janeiro, June 1992), (Ahmedabad, India: Centre for Environment Education, 1992) p. 32.

21.   Commission on Developing Countries and Global Change, *For Earth's Sake* (Ottawa: International Development Research Centre, 1992) p. 37.

22.   *Tribune* (Chandigarh, India) 13 June 1992.

23.   M.V. Naidu, 'Ideology, Technology and Economic Development,' in Dwivedi (ed.), *Perspectives on Technology & Development*, p. 30.

24.   See Richard Appelbaum, *Theories of Social Change* (Chicago: Markham Publishing, 1971) pp. 15–17.

25.   UNESCO, *Moving Towards Change* (Paris: UNESCO, 1976) p. 70.

26.   See Willy Brandt (Chairman), The Report of the Independent Commission on International Development Issues, *North–South: A Program for Survival* (Cambridge, Mass.: MIT Press, 1980).

27.   See, for example, Peter Bowler, 'Will Science and Technology bring Conflict within Third World Cultures?', *Science Forum*, Vol. 10, No. 3, June 1977, pp. 12–13.

28.   M.V. Naidu, 'Ideology, Technology and Economic Development', pp. 11–33.

5 Environment and Development

INTRODUCTION

Since the Second World War, the world has faced two major crises of global proportion (not counting the threat of nuclear war). One was the energy crisis of the 1970s which forced the industrialised nations to take various steps to reduce the possibility of a recurrence of such a crisis. One by-product of this crisis was NIEO, which has been referred to in Chapter 1. The other, potentially calamitous, crisis is the survival of our planet earth. It is the environmental crisis which has brought people of the North and South together because they are all concerned about the fate of their earth (although they may differ about the plan of action to save it).

Governments, international aid agencies and the general public all over the world are now worried about this crisis. Unlike the previous energy crisis, which affected the citizens of the industrialised countries much more than those of the poor nations of the South, this time the very resource base of humanity is threatened, and it is on this base that humanity's present livelihood as well as its future depends. This is equally relevant even to those who live in a state of dire and debilitating poverty on a daily basis. The challenge before the governments of developing nations is to find a way of marshalling their limited financial resources towards improving the quality of life of their people. This crisis has further complicated the issue of administration for development because in addition to continuing problems with respect to economic planning, industrial and agricultural productivity, energy production, transportation and communications, and the general welfare of people, a host of other obstacles has emerged recently. These include:

(1) the growing human population, which is making increasing demands on natural resources thereby reducing resources in an unsustainable manner;

(2) the indifference of industries (both in the public and private sectors) to environmental protection and the safety of workers, causing pollution of air and water as well as degrading other aspects of the environment;

(3) the continuation of various development projects (supported until very recently by international aid agencies) which were conceived without considering their impact on ecology and society;

(4) the ineffectiveness of government environmental protection policies, programmes and laws, with widespread laxity in monitoring, enforcement and prosecution;

(5) the gross undervaluation by the general public of the importance of protecting the environment.[1]

The strange thing is that in all such cases, it is the poor and the marginalised who suffer most. Therefore, unless those who administer developmental programmes accept and are sensitive to the necessity of linking developmental programmes with environmental conservation and protection, in time they will no longer be able to sustain them. Thus, the governments of developing countries have no choice but to adopt national sustainable and environmentally sound development strategies. What these strategies and institutional frameworks ought to be, what type of policies and action plans should be considered by these governments, and how the administrators should be trained for environmentally sound and sustainable development (ESSD) policies and programmes, will be discussed in this chapter.

ENVIRONMENTALLY SOUND AND SUSTAINABLE DEVELOPMENT (ESSD)

The concept, environmentally sound and sustainable development, envisions a policy stand based on the government's realisation that long-term economic growth is only possible if a nation sustains its viable natural environment and adopts an integrated approach to sustainable development. It recognises the interconnectedness between scientific, technological, social, cultural and economic dimensions. Further, such a country must believe that its people have the right to live in ecological harmony. That is why the environmental policy of each nation should be based on ESSD. It is this policy stand which may engender the necessary interconnection and co-ordination between various policy sectors in that government, such as fulfilling basic needs and conserving the natural resource base.

This requires, however, a major change in the policy perspective of a government so that environmental considerations become an intrinsic part of national policy-making and are not added afterwards. And among the first steps to be taken is the determination to act on the local environmental health issues rather than worrying about the international resource and environmental problems. For the developing nations, protecting the health of their people and conserving their natural assets is the most important policy action. Preventing adverse environmental impact is better and cheaper than firefighting action after the disaster has taken place. That is why environmental values in decision-making should be woven into the policy framework by the governments of Third World nations. These values stretch from bio-diversity protection and conservation, sustainable resource use and environmental protection, to fostering economic development of a kind that both harmonises with the natural environment and augments the quality of life for all the people, present and future.

Sustainable Development

Development or economic progress has always produced, as a side effect, the degradation of the environment. The more the material progress in general, the more the deterioration. This has been an integral part of the developmental process so far because we have hitherto taken a compartmentalised view of the situation rather than taking a holistic perspective on development. This latter multi-sectoral approach to the developmental paradigm is of recent origin, but it is the type of panoramic view that is necessary if we are to maintain sustainable development. Seen from this perspective, economic growth has to be intertwined with the question of environmental equilibrium, maintenance of inter-generational equity, and minimisation of natural resources depletion. These elements need not be at cross purposes, but rather may be complementary.

Sustainable development as a term was advocated by the World Commission on Environment and Development. Since 1987 when the commission brought out its report, *Our Common Future*, this term has united people around the earth over the need to save our planet and to create an environmental stewardship. The commission defined the term as: 'meeting the needs of the present generation without compromising the needs of future generations'.[2] This definition has been endorsed by the World Bank in its 1992 World Development Report, *Development and the Environment*:

> Sustainable development is development that lasts. A specific concern is that those who enjoy the fruits of economic development today may be making future generations worse off by excessively degrading the earth's resources and polluting the earth's environment.[3]

Although that definition is not sufficiently comprehensive, nevertheless, for the first time, the World Bank has acknowledged that 'meeting the needs of the poor in this generation is an essential aspect of sustainably meeting the needs of subsequent generations'.[4] Hence, the previous dichotomy between economic development and environmental conservation and protection has been neutralized. Even so, the World Bank is using a narrower definition by emphasising benefits and costs and macroeconomic analysis to strengthen environmental protection in order to obtain rising and sustainable levels of welfare.

A framework for sustainable development may have to vary from country to country. For example, in the case of countries such as India and China it has to be charted against the backdrop of tremendous population pressure, which in turn has given rise to deforestation, soil erosion, the silting up of rivers and streams, and desertification. Development for such nations as well as for others will have to become synonymous with environmental protection and conservation. Sustainable social and economic development calls for a determined fight against poverty, which is related to population pressure. Thus, at the national level, developing nations should consider creating the framework of a national population policy which recognises the interaction between population and the environment. This framework would call for action on three fronts: evaluating the environmental implications of population growth and its urban–rural mix; assessing the environmental impact of the use by the public of natural resources; and considering human-centred development measures (such as fulfilling basic needs and upgrading the status of women) as an integral part of developmental policies.

It should be noted that the state of the environment of a nation cannot be isolated from the state of the world environment and economic development. It is a closed circle. Developing nations are acutely aware of the fact that poverty is the greatest threat to the quality of the environment anywhere. The poor not only suffer extensively from the environmental damage caused by their rich cousins, they themselves are the cause of environmental decline in their own countries. Environmental degradation and economic deprivation are interrelated. No country or group of countries can tackle the world-wide problem of environmental pollution single-handedly. We live in a world of shared and interacting environmental resources. It is imperative that we all believe that a clean environment is the right of all people anywhere on earth.

AN INSTITUTIONAL FRAMEWORK FOR PROTECTING THE ENVIRONMENT

The continuing and accelerating deterioration of our planet's ecological base poses a major threat to the viability of our world and nowhere is the evidence of global ecological deterioration better argued than in *Our Common Future*, a Report by the World Commission on Environment and Development (WCED). But the real significance of this report lies in its thorough explanation of why the people inhabiting this planet are collectively destroying the resources. Its thesis is that we cannot save the environment without development, but at the same time we cannot keep on developing unless we save the environment. Yet without restructuring the existing institutional arrangements and legal mechanisms, fragmentation, overlapping jurisdictions, narrow mandates and closed decision-making processes will continue. In its main recommendation the report states that as 'the real world of interlocked economy and ecological systems will not change, the policies and institutions concerned must'. To change the existing institutions and legal mechanisms, the commission suggested that nation states should consider one of two approaches: either establishing an 'environmental policy, laws and institutions that focus on environmental effects', or adopting 'an approach concentrating on the policies that are the sources of those effects'.[5] Both approaches represent a distinct way of looking at solving environmental issues.

Most national governments have adopted the first approach – effect-oriented – by establishing environmental protection agencies without much change in existing institutional structures. Such agencies have been given jurisdictions mostly in those areas to which no other ministry or department could lay a concrete claim. However, as experience in some industrialised countries has demonstrated, this approach has resulted in focusing exclusively on the effects and on reactive or remedial actions, rather than on preventative measures. This has created some serious jurisdictional disputes concerning the environment, which is seen by many ministries as a shared responsibility. In some cases, the environmental protection ministry or agency has taken a sharp stand against economic development policies being pursued by others. Thus, inter-ministerial jealousies have resulted in jurisdictional fights, policy fragmentation, and pursuance of narrow objectives. The need of the day is not only to dwell on the effects of environmental pollution but also on its sources and causes. It is essential that all government ministries, whether dealing with economic growth, natural resources, human health, social development or any aspect of human endeavour are mandated to care for the environment. That is why the following recommendation of the WCED is the key to the basic assertion of this chapter: the necessary interconnectedness among all governmental institutions:

> Environmental protection and sustainable development must be an integral part of the mandates of all agencies of governments, of international organizations, and of major private-sector institutions. These must be made responsible and accountable for ensuring that their policies, programmes, and budgets encourage and support activities that are economically and ecologically sustainable both in the short and longer terms.[6]

Nowhere would a real environmentalist argue against economic development if it is done in an ecologically sound and sustainable manner. This can be achieved if each country, to the extent that it is capable on its own, puts its house in order by bringing necessary changes in its institutions and in its attitudes towards nature. This is an obligation that each country has towards its citizens and to the world at large. Thus each country, irrespective of its size and economic status, must set a good example at home before it demands that others respond to the global ecological crisis.

A POLICY FRAMEWORK FOR ENVIRONMENTAL PROTECTION

The issue of environmental protection has gained priority on the political agenda of many developing nations. Public awareness about the environment has succeeded in putting this issue at the centre of decision-making in government, industry and the home. During the 1990s, but especially since the Earth Summit in Brazil, public concern has become strong enough to compel a change in the political rhetoric of all nations. However it has not yet become strong enough to force real changes in the kinds of decisions that government and industry have always made – or avoided making. Policies on tax and fiscal incentives on energy, agriculture and forestry have not yet changed significantly. These continue to cause a steady depletion of the basic stock of ecological capital. In many Third World nations, the debate has begun only recently on the policy and institutional changes needed to reverse environmental degradation and move towards sustainable development; obviously, it is going to take time.

Each nation needs to bring the environment from the margins to the mainstream of political thinking and decision making. This requires establishing a national policy framework for environmental protection and management by each country. The framework (see Figure 5.1) illustrates several interlocking mechanisms which should be considered by a nation in seeking to develop an integrated environmental policy and pollution control strategy. Such a framework should be based on certain approaches which are listed below, and may be used as a model which Third World countries can adopt to tackle the environmental and natural resource challenges facing them.

(a) *Policy Instruments*

A country can utilise two types of policy instruments to attain its pollution control objectives: a regulatory mechanism, or an effluent charge strategy. The *regulatory mechanism* requires that the authorities take the following four steps: set rules and regulations governing

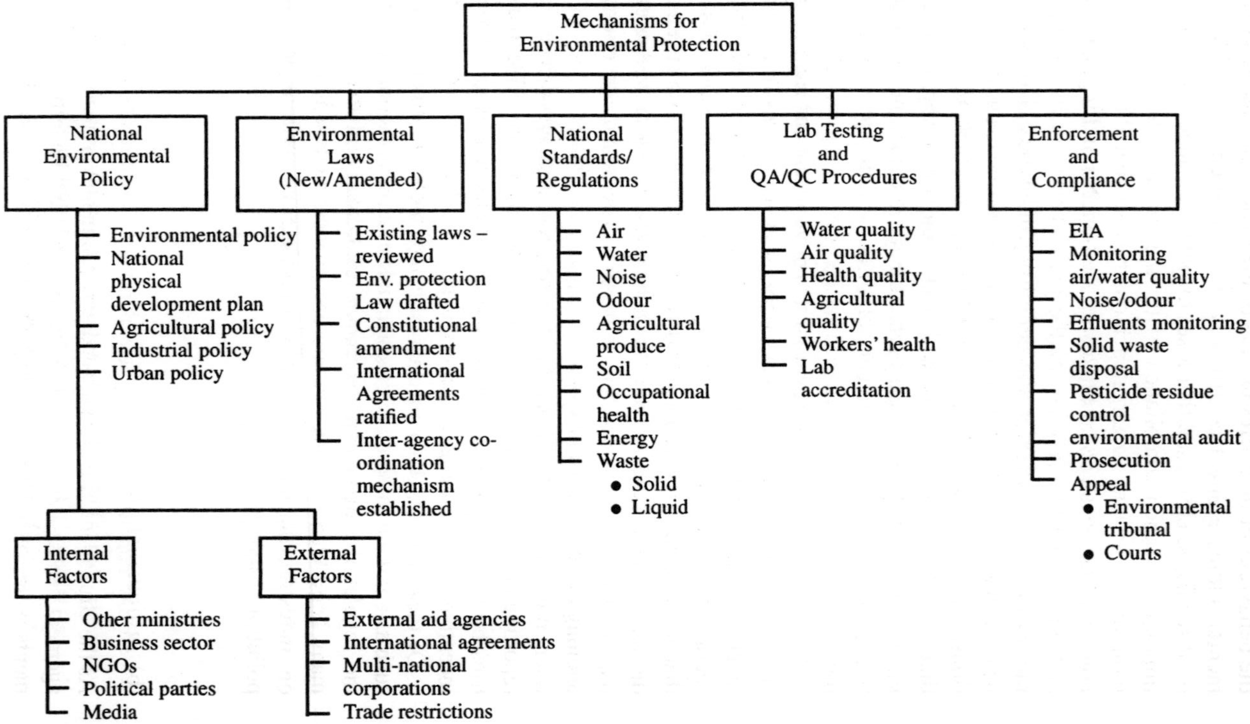

Figure 5.1   National framework for environmental policy and management

the behaviour of industries in each of the sectors (including environmental audits, effluent control devices, maximum allowable limits on discharges, etc); establish a set of penalties under the law to be imposed for non-compliance with the regulations; continually monitor the actions of targeted industries so that the instances of non-compliance can be detected with spot-checks and regular audits; and finally make timely use of the judicial process in seeking the imposition of penalties on the defaulting industries. The effectiveness of this approach is based on the premise that even minor violations, if detected, will not be ignored by the enforcement authorities; that defaulting industries should first be given an opportunity to mend their ways before being forced to pay major fines or injunctions; and that the enforcement authority has sufficient budget and human resources to enforce the law.

The *effluent charge strategy* uses economic incentives such as effluent or emission charges. In order for this strategy to work, the enforcement authority ought to consider the following three steps: determine a set of charges or prices per unit of discharge of each polluting substance that is predicted to induce the necessary abatement actions on the part of dischargers; continually monitor the level of discharges as well as establish a system of self-reporting with spot checks and environmental auditing mechanisms; and levy a sum equal to the charge per unit of pollutant multiplied by the amount of the pollutant discharged during each reporting period. This approach provides a graduated incentive to industries by making pollution itself a cost of production; it also provides incentives for technological innovations.

Of these two basic approaches (although there are other variations such as a mixture of the two), most countries have followed, in general, the regulation-enforcement technique. Particularly in the context of many Third World nations, this approach is more successful than relying on industries to innovate and self-regulate as well as to control pollution.

(b) *Institutional Change*

As the Brundtland Report indicated, environmental protection and management can no longer be safely left to weak and underfunded departments or to several departments with overlapping responsibilities. The issue of integrated pollution control must be addressed responsibly; towards this end, all nations (especially the developing countries, if they have not already done so) have established departments/ministries of environment as a structural response to the changing perceptions of the environmental challenge.

(c) *Legislation Needed*

To protect public health and the environment, effective laws should be passed and vigorously enforced. Legislative and regulatory instruments set rules and regulations for the protection of the environment. For example, through the Environmental Impact Assessment (EIA) process, government should plan to ensure that environmental factors are considered in decision-making. This can be done through prohibition, standards, guidelines, permits and the like if each government is committed to making its enforcement capability highly effective.

(d) *National Environmental Standards*

National standards pertaining to water, air, effluent emission, noise, waste, pesticide residues and odour should be established by all governments. It should be noted though that strengthened standards can only be achieved through increased investments from both public and private sectors as well as by concerted action on the part of the government regulatory agency.

(e) *Laboratory Testing and QA/QC Procedures*

Constantly to monitor and detect the quality of industrial emissions, effluents, and other polluting substances (both organic and inorganic), a nation needs a network of environmental laboratories (in public as well as in private sectors). To ensure that the procedures used by all the labs (specially those in the private sector, or even those operated by other ministries) for quantification are appropriate, and whether properly calibrated analytical instruments are being used, there should be a proper accreditation programme to maintain the necessary quality assurance (QA) and quality control (QC) procedures.[7] Without such a programme, prosecution of an environmentally-related offence may be difficult if the competency of the laboratory doing testing and analysis is in question.

(f) *Environmental Enforcement and Compliance*

Securing compliance with environmental legislation and regulation is a mandatory activity for the governmental environmental department. The success of an enforcement and compliance strategy will depend on gathering sufficient evidence and information on the suspected violation. The evidence in the form of a sample is generally sent to an accredited laboratory for analysis and testing. Enforcement involves regular inspection and monitoring to verify compliance, investigation of any violations, and applying measures to compel compliance. Prosecution and conviction is the final stage in the compliance strategy.

In addition to these main approaches, the following six items are equally important tools, developing a sound policy process and establishing institutional mechanisms for environmental protection.

(1) *Long-term Targets*

The traditional incremental approach to environmental protection needs to be replaced by a clearly defined long-term target-led strategy. For example, the ban of coral-sand mining in coastal waters and its replacement by basaltic sand within a time-frame of three years illustrates such a long term objective on a delicate issue. The same may be said of the CFCs protocol to the Ozone Convention and the reduction of CO_2 emissions, albeit on a global scale.

(2) *Integrated Pollution Control (IPC)*

One of the toughest problems in dealing with the environment is the degree to which it is an integrated system – a delicate whole in which a change to one part affects all the others. This makes it difficult to set priorities for environmental action. Environmental problems do not easily lend themselves to this kind of ranking. If, for example, water pollution were to be the first issue tackled, what would happen to initiatives to deal with air pollution, which is itself a primary source of pollutants in water?

(3) *Public Participation in Decision-making*

There is a growing recognition of the fact that the public has a right to information and that more direct citizen involvement will reduce conflict, enhance trust in agency decisions, and improve the quality of decision-making. Towards this, public input into decisions should be encouraged. This involvement will lead to substantive contributions to environmental protection by raising new issues and serving to counterbalance narrow agency biases. Active participation is necessary to ensure that policies reflect public preferences, and procedural justice is necessary to foster public acceptance of government decisions.

(4) *Economic Instruments*

Economic instruments that reflect environmental costs will encourage decision-makers to take the environmental consequences of their actions into account. Possible measures include effluent taxes, tradable emission rights, deposit/refund systems and user charges. Sustainable development, with its focus on anticipating and preventing environmental degradation before it occurs, means increased emphasis on appropriate resource pricing and economic instruments to achieve environmental objectives. Used properly, these can ensure that the environment is more fully considered in production and consumption decisions made at all levels of society.

(5) *Changing Decision-making Processes*

It is not enough to improve the quality and availability of factors that affect decision-making. The existing institutions and procedures must also be changed. Environmental considerations must be formally recognised as essential decision-making criteria within government and private sector organisations. Improved decision-making, which takes full account of environmental considerations, demands that all these partnerships be strengthened and expanded.

(6) *Strengthening Partnerships*

Better environmental decision-making will require co-operative efforts at all levels. There are many other important partnerships that need to be fostered. *Business* is an essential partner in the search for, and implementation of, effective solutions to environmental problems. *Labour* has an important role in changing the way decisions are made and in working with governments and the business community to achieve environmental objectives. *Women*, individually and through the many organisations of which they are a part, are also a key to changed decision-making. These partnerships will be essential to the long-term goal of achieving sustainable development. The *environmental non-governmental organizations* (ENGOs) are playing a crucial role in educating people about environmental issues and environmental activity. They are essential to an active and balanced discussion of environmental issues.

Individuals and communities have the most important role to play in bringing about changes in our attitudes towards the environment and how we treat it. Towards this end, governments should play an important role in encouraging individual and community actions in environmental conservation. A worthwhile idea is to set up a National Environment Fund which could provide a springboard for co-operative government–community efforts to protect and preserve the environment; it might also assist in the promotion of environmental education and research, in the support of ENGOs, and in encouraging locally-initiated environmental programmes.

Today's *youth* are tomorrow's decision-makers. If they are to play a meaningful role in environmental decision-making and diplomacy, they must develop environmental literacy as well as leadership skills. There is a need to take measures to ensure that the nation's youth will have the opportunity to channel their energy and talents towards achieving environmental objectives.

At the international level, global problems require global solutions. Such solutions can be effective only if they are built on *international partnerships*, both bilateral and multilateral. This will require raising a collective consciousness among all the nations so that a concrete shape to the slogan 'think globally and act locally' can be given.

This framework should be supplemented by an action plan which translates these perspectives into concrete legislative, policy and institutional systems.[8]

ACTION PLAN FOR ESSD

Today's environmental problems are the result of the failure of people at all levels of society to make decisions that fully take the environment into account. This failure, especially in the developing nations, is not one of wilful irresponsibility. Indeed, it is only recently that these people have started expressing their concern about environmental issues. Now the challenge is to integrate environmental considerations into decision-making in a more systematic, focused and co-ordinated way. Therefore, for the governments of developing nations, their first priority should be to help change the way certain of their policies are formulated, institutions operate, and decisions are made. In particular, they should consider establishing a national environmental investment programme which provides goals, objectives, directions and guidance for governmental action with explicit linkage between environment and sustainable development. An action plan or framework should be drawn up which should include at least the following four policy areas for immediate action: a national physical development plan for the country; a national environmental policy along with the relevant administrative mechanisms; comprehensive legislation for environmental protection and management; and an appropriate industrial policy. These aspects are discussed below.

A National Physical Development Plan

Many of the environmental stresses specially pertaining to urban development and siting of industrial complexes result from ambiguous planning. Orderly and progressive development of land, towns and other areas, whether urban or rural, is a prerequisite for sound environmental planning. Towards this end, each of the developing nations should consider preparing a National Physical Development Plan (NPDP) for itself. Such a plan should contain an assessment of development priorities with associated economic and financial costs, coupled with the delineation of an appropriate institutional framework for the implementation of the plan. An NPDP should be the first step in creating a sound environmental protection strategy for any nation. In the absence of an NPDP, there would emerge inefficiency in land management, leading to further environmental degradation. The NPDP should consist of a map of the national territory, with supporting plans and documents. These would indicate the preferred spatial strategy, after reviewing all physical, social, historical, economic, environmental, political as well as international issues, and after taking into account fundamental goals for the nation's future spatial organization.

The main objectives of the NPDP may include:

(1) the protection of agricultural areas, especially areas with the best potential;

(2) the protection of the natural heritage, including the terrestrial and marine ecosystems, the coastline, historic and religious buildings and areas;

(3) the judicious siting of all secondary and tertiary activities, taking into consideration socio-economic as well as environmental issues;

(4) the improvement in the quality of life of all its citizens. This would include all the spatial aspects related to the educational, employment, social and leisure needs of the community, as well as the transportation and communication sectors.

Based on the NPDP, or even when it is being prepared, each government should consider preparing a national environmental policy, because these two policy instruments together will assist the government in formulating an appropriate institutional strategy.

A National Environmental Policy and Institutions

It is suggested that the main goal of a national environmental policy should be to foster harmony between quality of life, environmental protection and sustainable development for present and future generations. In addition, the policy should include the following three specific features: [9]

Guiding Principles

These include conservation and the preservation of the environment, accepting the stewardship role for mother earth, taking due care in planning economic development so that future generations will not be unduly harmed, undertaking preventive and anticipatory actions to control environmental pollution, recognising the need for and importance of collective decision-making processes in environmental planning and management, providing for appropriate legislation so as to establish adequate enforcement mechanisms, acknowledging the impact of international environmental issues on each state and their role in minimizing international pollution, promoting environmental codes of conduct, and encouraging environmental education and awareness.

Main Objectives

Some specific objectives should be agreed upon so that the state is directed to act on such items as attaining industrial development while conserving the nation's natural resources base, providing access by all

people to amenities such as clean water and sewerage, promoting environmentally sound technologies, safeguarding occupational health and
safety of workers, conserving and preserving the natural and cultural
heritage of the nation, establishing and enforcing air and water quality
standards, and co-operating with regional and international organisations
for the global protection of the environment.

Specific Attributes of the Policy

In addition to the main objectives, the policy should have some specific
characteristics. These may include: (1) meeting basic human needs without endangering the environment; (2) utilising natural resources in a manner which is ecologically efficient; (3) requiring environmental impact
assessment of any major development project; (4) using economic instruments in achieving cost-effective environmental management; (5) controlling pollution at source and, to the extent possible, requiring the
polluter to pay; (6) undertaking by government to provide clean water to
all residents; (7) preparing and implementing a national master plan for
waste management; (8) in order to reduce air pollution from exhaust,
requiring vehicles to have exhaust control devices and encouraging the
use of unleaded petrol; (9) controlling noise pollution; (10) creating green
spaces around the urban settings and establishing national parks and sanc-
tuaries for wildlife; (11) exploring any means for energy generation by
recycling household and industrial wastes, and establishing a recycling
mechanism; (12) issuing nation-wide air and water quality standards,
guidelines and codes of practice for use by all industrial and commercial
concerns; (13) undertaking a qualitative study to assess pesticide residue
in soils, foodstuffs, fish and animals; (14) promoting environmental education at all levels of schooling, promoting research on the causes and
effects of environmentally related diseases, assisting NGOs and cultural
groups in enhancing environmental awareness and assisting in the training
of scientific and technical personnel needed by the country; (15) introducing comprehensive environmental protection legislation; (16) prohibiting
smoking in all designated public places; and (17) establishing a national
fund for the environment.

These suggested policy guidelines require some specific institutional
and administrative mechanisms in order to implement, enforce and
manage the recommended course of action as envisaged in the
suggested policy document. Most important among these are:

(1) establishment of a Department of the Environment to be responsible for the administration of the environmental protection legislation, to design and develop environmental guidelines and standards, and to be a national focal point for information and research on all environmental matters;

(2) incorporation of environmental considerations in the National Physical Development Plan;

(3) establishment of an environmental council representing industries, unions, NGOs, educational and research institutions, professional associations, cultural and religious groups, employers' federations, media, and others;

(4) provision of a high-level central co-ordination and policy-approval mechanism to act as a pivotal body to ensure that the wishes of the legislature are carried out by all government ministries and parastatal organisations;

(5) a collaborative mechanism to be put in place to seek co-operation from the provincial, regional and municipal governments in areas of shared responsibilities and common concern – for recreation, conservation and protection of the environment, management of land-fill sites, garbage removal, crisis management, enforcement of environmental laws and regulations;

(6) a requirement for environmental impact assessments on all developmental and industrial projects;

(7) encouragement of media (television, newspapers, and radio) to be partners in enhancing environmental awareness in the country;

(8) establishment of an integrated and comprehensive national system for environmental protection and conservation.

The Legal Context

A discussion about the institutional arrangements for environmental protection is incomplete without also addressing the legal context of those institutions which are in existence or to be created under the enabling legislation. Some difficulties may be encountered initially because the combination of government institutions and legal arrangements may be very cumbersome. I am using the term 'cumbersome' here because the laws, rules and regulations, orders, ministries, departments, divisions, para-statal bodies, commissions, boards, and councils create a complex web of interrelationships which perplex a citizen.

Nevertheless, it is essential that a review of the existing legal situation should be undertaken to determine the scope and extent of jurisdictional authority, the existing gap, and remedial action suggested.[10]

It is not suggested here that all pollution control jurisdictions be centralised into one ministry or a department. Decentralisation is sometimes better than a highly centralised organisation. This is the case when a particular agency specialises in one sector of the environment. The best examples of this are the areas of fisheries and hazardous substances. In many countries, the decision has been made to have fisheries departments, rather than environmental protection departments, deal with marine pollution. The reason for this is that, prior to sophisticated analytical methods, fish were the best indicators of pollution: they either died or left polluted areas. Fisheries officers and the fishermen they deal with on a daily basis can be the best marine pollution policemen, since they know local fishing areas, the movement of local waters and the behaviour of marine life in the area, and are the first to be aware when something is wrong.

For most of the nations, centralisation of all forms of pollution control is neither necessary nor even desirable; nevertheless, the decentralized system requires that where several ministries are involved, there must be some means of ensuring that the various environmental efforts are co-ordinated and do not duplicate, or even worse, contradict each other. Having two ministries perform the same or similar functions is not an efficient use of government resources and the taxpayer's money, and it also confuses the public as to which standards or rules to follow. Duplication also creates legal problems, since a person cannot be convicted twice for the same act or omission even though the offence may have been committed under two separate legislations administered by two ministries. Obviously, to exert the authority necessary to arbitrate between competing programmes of different ministries, the co-ordinating body has to be a powerful agency, most likely to be headed by either the prime minister or the president.

For securing an appropriate legal base, three options are suggested:

(1) The first option is to leave responsibility for environmental protection and management to the existing ministries but to strengthen the present laws. This option involves the least change in the administration of the government. However, as environmental pollution control is generally fragmented among several ministries, the situ-

ation becomes more complicated and duplication as well as jurisdictional disputes get perpetuated.

(2) The second option is to opt for comprehensive environmental legislation which would create a powerful central agency with overall responsibility for the environment. Any legislation which duplicates or conflicts with the new law would have to be repealed. In effect, this option would create a super environmental agency or ministry. This would mean radical changes in terms of the organisation of the government and the administration and distribution of power among various ministries. However, this option may be more easily implemented and more effective in a small nation rather than in a large federal state.

(3) The third option is a compromise between the first two. It involves the introduction of a comprehensive environmental law, with the power entrusted to an environmental ministry to set national standards on air and water quality and to be given overall control and management of the environment. The new law may cover all of the gaps in environmental protection, and also permit ministries with existing environmental responsibilities to carry on and even strengthen their powers in their particular jurisdiction. The environmental law performs a 'backstopping' as well as 'audit' function for other ministries with existing responsibilities, but would operate as the primary law for those areas of the environment that have no coverage. This option brings about relatively minor changes in governmental organisation and structure, but requires an integrated system of co-ordination so that the potential for duplication of effort and enforcement is minimised. This option also entails a much more active role by the central environmental policy agency (as suggested earlier) as the final arbiter of sectoral and jurisdictional differences.

There are, in general, two approaches to the enforcement of environmental laws: the positive enforcement approach, and the punitive approach. The positive approach is envisioned to provide corrective measures such as programme approval, enforcement and prohibition notices. The punitive approach seeks to go to the extreme of imposing the legal sanctions of the law. This authority is exercised by the courts with appropriate jurisdiction when criminal and civil aspects are involved. The recent origin of environmental law and the consequent necessity to define new terms and concepts and to redefine old ones justifies greater judicial participation. It is for these reasons that, instead of using administrative courts, it would be desirable to establish quasi-judicial bodies such as an environment tribunal. Such a tribunal should have the powers to review the action of the government's environmental ministry and to see that it functions in a proper manner. Further, the penal sanction in environmental legislation should include a considerable range of fines as well as imprisonment. To be effective, fines must be tough enough to have a deterrent effect on the target industry. If the fines are moderate or low, they can easily be included in the industry overheads and passed on to the consumer. Thus it creates a double burden on the public, of having to bear the effects of pollution and of having to pay for it.

An Industrial and Environmental Policy

The promotion of environmental protection techniques and low-waste technology is an essential preventive instrument in environmental policy, supporting efforts towards sustainable industrial development. Consequently, appropriate fiscal or monetary policies have to be formulated to encourage economical use through easy accessibility to soft loans. A consumer policy also has to be developed. The complexity of environmental choices for the consumer is confusing. The relentless increase in consumption contributes substantially to the pollution of affluence. This policy should focus on ways in which the consumption of products detrimental to the energy economy and harmful to the environment can be regulated through taxes and charges as well as by restricting the use and production of disposable products apt to despoil the environment. Both economic and ecological considerations deserve to be taken into account in industrial decision-making. An increase in the efficiency of energy and materials, for instance, would meet ecological objectives and also cut down on costs.

Countries lagging behind will find themselves eventually compelled to choose between home markets and settling for a lower quality of environment, or importing technologies developed by greener competitors.[11] During the remaining years of the 1990s, the approach to environmental issues looks likely to be governed by considerations of

(1) adopting practicable methods for controlling waste disposal right from the time of project formulation;
(2) adopting clean technologies and encouraging more effective re-use and recycling of waste in production;

(3) implementing a toxic and hazardous wastes management programme;
(4) implementing demonstration projects in the control of industrial pollution with emphasis on clean technologies in specific industries, resource recovery, waste recycling and utilisation, as well as proper waste disposal.

Industrialisation transforms raw materials into consumer products. Badly managed firms are rarely kind to the environment. By contrast, efficiently-run firms try hardest to reduce environment damage. The risk of the manufacturing activities despoiling the environment, wasting raw materials or dumping wastes haphazardly grows proportionately with the rate of industrialisation and the absence of resource management. Fortunately, technology is available to mitigate these trends. But the cost associated with environmental degradation rarely enters into our calculations. The time has come to add such costs as health care and the protection of nature. We need to improve our accounting system and modify our GDP calculation, for instance, to reflect the new cost of economic activity. Health protection, water purification, or the conservation of nature reserves and parks are taken to have a zero price in the absence of a marketplace mechanism to determine their real values. This approach should be changed to integrate the environment variable into economic planning.

The above suggested plan for action should be studied carefully by the policy-makers of the developing nations so that each country has its own unique set of policy and programmes created to strengthen institutions for environmental protection and management. But the effectiveness of this framework depends largely upon how the various governmental ministries and agencies co-operate in implementing the environmental policy and programmes developed for their respective nations. This aspect of interconnectedness is examined in the next section.

THE JURISDICTIONAL ASPECT OF INSTITUTIONAL ARRANGEMENTS FOR ENVIRONMENTAL ADMINISTRATION

'Institutional arrangements' generally means a set of rule-ordered relationships among organisations created by human beings to achieve a set of objectives. The important thing to note here is that any one institution or organisation, acting alone, could not accomplish all of those objectives. Team work, joint activity and the use of resources (human, financial and physical) in a co-ordinated manner is crucial whenever more than one person or institution are involved. It should also be understood that institutions are the means by which people order their relationships with others for productive, collaborative and compatible arrangements. That is why the structures of institutional arrangements are critical in our understanding of how they interact with each other, and how they enhance certain capabilities and impose certain limitations.

One more factor needs to be elaborated here. In the western world, there is a frequent assumption that people can accomplish anything if only they have the will and determination. However, will and determination alone are inadequate in dealing with already established organisations and institutions of collective action. How true this is in the case of the environment or the 'tragedy of the commons'! In any country where this factor is not taken into consideration in planning for environmental institutions, jurisdictional problems and organisational frustrations have arisen later. That is why the intersectoral relationship must be strengthened, to be followed by tight inter-agency co-ordinating mechanisms. These are highlighted below.

Intersectoral Aspects of Environmental Administration

There is no doubt that our existing institutions for environmental protection are unsuited to new tasks. For example, old public health criteria are no longer sufficient for present-day circumstances; hence the job cannot be done unilaterally by the ministry of health alone. Similarly, an array of single-purpose agencies each being responsible for one resource, such as water or agriculture, cannot be expected voluntarily to subordinate or relinquish their specific mandates to a vague goal such as enhancing environmental quality or the quality of life. Further, it is not easy for the older institutions simply to hand over their interests to a new institution whose jurisdiction concerning environmental protection cuts across the existing administrative and jurisdictional boundaries. Moreover, the preference of some that all aspects of the environment ought to be handled by a super ministry is equally troublesome to the existing governing institutions. Hence, the solution lies in adopting a realistic approach based on compromise and mutual adjustment among the older ministries and the newcomers. The sooner this realization gets accepted, the better it is for the environment and for the community. The central theme that should determine the task of environmental protection is the prevention and control of factors or situations in advance of crises. If preventive measures are taken, the community and the nation can be spared possible disasters. Treatment or other after-the-fact remedies should be regarded by the governmental agencies as a failure to achieve this objective.

The Concept of Enforcing Agency

In some nations, the jurisdiction for environmental conservation and protection is a shared responsibility for special matters such as air quality, water quality, noise and so on. There is need to have a provision whereby standards, regulations, and the enforcement provisions of the law may be applied by designated enforcement agencies. These agencies could continue to be responsible (if that is what they were doing in the past) for monitoring air, noise, quality control of water for drinking and domestic purposes, odour, effluents, solid wastes and pesticide residue. They can be given authority pertaining to programme approval, enforcement notice, variation notice, and the necessary power to secure compliance with the law.

Inter-ministerial Coordination for Integrated Pollution Control

Due to the multi-disciplinary nature of environmental issues and their management, all governmental agencies involved in environmental protection must co-ordinate their activities with their sister ministries at different hierarchical levels. For this, an integrated pollution control (IPC) system should be adopted which can establish linkages among relevant ministries in order to avoid a fragmented managerial approach to environmental protection. Integration of environmental protection considerations requires both vertical and horizontal linkages. Both of these approaches should be used by the government.

The vertical linkage means that the country has an IPC mechanism, with a common theme linking such items as nuisances (for example, noise), clean air, clean water, control of hazardous substances (including pesticides, radio-active materials), litter, solid waste collection and disposal, sewage treatment, dumping at sea, exhaust gases from motor vehicles, obnoxious smells, dust and ashes, environmental health, natural resources, historical and cultural properties, and other commons. Such a system is difficult to establish, even in some unitary states, because some of these sectors traditionally have been managed by various government ministries. Nations rarely demolish their existing governing instruments, certainly not for an amorphous subject like the environment. The best one can hope for is to see a well-working horizontal co-ordinating approach.

Horizontal integration is possible through various means. Four such means are: through a high-powered inter-ministerial committee of cabinet (in a parliamentary form of government or a suitable body in other forms of government); through an inter-ministerial co-ordination committee for environmental protection; through inter-ministerial review and an evaluation of environmental impact assessment statements submitted by proponents; and on a technical side, scientific integration through environmental labs (such as an agriculture ministry lab, a health ministry lab, and the lab of the environmental ministry) for the monitoring of contaminants in the environment. Among these four approaches, the establishment of a central environmental policy co-ordination agency is crucial; and in order for this agency to have a standing among the various inter-ministerial bodies already in place, it should be given a legal status under the nation's environmental legislation.

The need for an inter-ministerial co-ordination committee can be demonstrated, as each country should have a strategy for environmental protection and management which requires a necessary interconnection and co-ordination between the various policy fields currently managed by different ministries. In several nations, compartmentalised thinking regarding environmental protection and conservation has generated fragmented decision-making; it needs to be replaced by multi-sectoral and ministerial co-ordination.[12] Such an inter-ministerial environment coordination committee could fulfill the following functions:

(1) ensure early and effective consultation on environmental protection;
(2) ensure a full and open sharing of information among the ministries and para-statal bodies on all matters related to the protection of the environment;
(3) advise the minister of environment and (when necessary) the national environmental apex policy co-ordination agency on matters related to environmental quality guidelines, national standards,

codes of practice, regulations, and other control measures in order to avoid duplication and to ensure proper enforcement to protect the environment;

(4) maximise co-operation and co-ordination and minimise conflict and duplication among ministries and parastatal bodies on environmental matters.

In the case of environmental impact assessment (EIA), a separate inter-ministerial consultation and co-ordination mechanism is required. There is also a need to establish an inter-ministerial technical committee to secure comments and suggestions on matters relating to environmental management and enforcement. Finally, concerning the interaction among various labs, a memorandum of agreement for testing procedures should be drawn up to avoid overlapping authority and jurisdiction among the labs.

The above suggested integrating mechanism needs to be discussed by those nations where such a system has yet to be put in place because an array of single-purpose ministries each managing a single resource such as water, agriculture, industry, air, or health, cannot be expected voluntarily to subordinate their specific mission/authority to a vague goal called environmental quality.

Environmental Crisis and Administrative Vision

Although concern for the environment has achieved considerable global acceptance and popularity, the perception of crisis is not shared by everyone in government and the bureaucracy. Those who articulate this sense of urgency are sometimes considered emotional and irrational power-grabbers and empire-builders. But the magnitude of environmental crisis demands a more measured, systematic and holistic response. Specifically, it demands institutional co-ordination, policy adjustments and structural adaptations. And when this demand has to be answered, existing governmental structures have no capacity to face so pervasive and complex a challenge and one as potentially disturbing to the administrative edifice of a nation. It is not surprising, therefore, to see individual ministries trying to protect their turf and to expand their power base by claiming that they have the administrative capability and vision necessary for dealing with environmental protection. In such an atmosphere, the task of co-ordination becomes extremely difficult, as the agency which has been entrusted with this task is seen by others as an undesirable intruder and obstructionist within their normal functions. The result can be futile gestures or withdrawal symptoms at worst, unenthusiastic support at best. Often, the resisting administrative agency employs such techniques as the prevention, curtailment or circumvention of anticipatory policy pronouncements. Finally, even when a new ministry of environment has been established, it does not do any better than being accorded a marginal position in the administrative hierarchy. Compounding the situation is the perception among many administrators that environmental policy and management is still a discrete function, something added on elsewhere to an administrative structure whose basic mandate has not been altered.

External Inducement for Institutional Capability Building

During the late 1980s many international aid agencies (such as the World Bank, UNEP, and northern nations' aid agencies) started showing great interest in a nation's capacity to manage its own environment. These external inducements helped to expedite the process of institutional acceptance and administrative accommodation relating to capacity building for environmental protection; nevertheless, the initial hesitation about the role of this 'new kid on the block' is still continuing. For example, this was clearly evident in India when the Department of Environment was formally established in 1980, or in Mauritius where the new Ministry of Environment and Quality of Life was constituted in 1991. It was not surprising to note that some officers of other ministries harboured concern about the new role of such an agency. Of course, this is nothing new because it has happened in other countries. Institutional strengthening takes a back seat when doubts (possibly to the point of outright resistance) persist among the key ministries about the new organization. In such a situation, direction from the office of the prime minister or president becomes crucial for ordering co-optation, integration, and co-ordination.

As environmental problems cut across ministerial responsibilities and jurisdictions, there is a need for more forceful articulation of environmental concerns in national planning and policy-making, as well as in improved co-ordination, institutional structures and legal instruments in the implementation of policies involving environment and development. For example, in the case of Mauritius, the World Bank, in its report

Economic Development with Environmental Management Strategies for Mauritius, noted that the country endures 'diluted authority and diffused legislation' which has implications for rational, efficient and sustainable natural resource management. Therefore, a concerted effort is required to systematically review the 'institutional, legal and regulatory arrangements, and the establishment of a comprehensive national policy for environmental protection and management'.[13] A similar situation exists in many other Third World countries.

A note of caution is desirable here. Externally induced changes, while easily instituted, must be given time to gain acceptance by the body politic and the existing administrative and legal mechanisms. Placing too much faith in instantaneous changes and expecting a 'magic wand remedy' for bringing requisite organisational effectiveness and motivational changes in the administrative apparatus have sometimes resulted in disappointment. And when the expected results do not follow, the recipient country turns against those organisations which induced such recommended changes. That is why great care should be taken before rushing towards instituting externally suggested changes. Like a human body, governmental institutions too require time to accept foreign organs; one should be ready for rejection from the body. It will be beneficial to all the Third World nations to undertake an assessment of induced change in the administrative, institutional and legal mechanisms pertaining to environmental protection.

Implementation and Action

Policy and legislation for environmental protection by themselves are not sufficient conditions for implementation and enforcement. These must be followed by the necessary institutional arrangements; but even institutional mechanisms must await the setting of priorities amongst the programmes which the government wishes to implement. Thus, after the policy has delineated areas for governmental action, and after legislative review and the framework have identified those areas which require legislative action, it is desirable for a government to set in motion a programme review process to determine which issues should be given priority, both in the near future and in the long run.

One major requirement should be for the government to establish a formal mechanism to be responsible for overall co-ordination and implementation of its policies and programmes. One such body suggested may be called an inter-ministerial environment coordination committee with the following objectives and functions: to ensure early and effective consultation on environmental protection; to ensure a full and open sharing of information among the ministries and parastatal bodies on all matters related to the protection of the environment; to advise the minister of the environment on matters related to environmental quality guidelines, national standards, codes of practice, regulations, and other control measures in order to avoid duplication and to ensure proper enforcement to protect the environment; and to maximise co-operation and co-ordination and minimise conflict and duplication among government ministries and parastatal bodies on activities dealing with environmental protection.

ACCOUNTABLE MANAGEMENT, CONSULTATION AND CO-OPTATION FOR PROTECTING THE ENVIRONMENT

Principles of Accountable Environmental Management

There are five essential principles for an accountable environmental management, which generally govern the relationship between government and two other major groups, the NGOs and the private sector.

Effective Application of Environmental Legislation

A clear understanding of all phases of the process is needed by all the participants, especially the NGOs. It is here where officers of a government ministry of environment should be extremely careful, and should avoid the temptation of cracking the whip too zealously since such an action may be counter-productive. For an environment officer to be too eager is to risk encouraging in polluters an un-co-operative attitude or forcing a legal battle.

Consultation

Throughout the process consultation is fundamental for sustaining public trust in the management of environmental policies and programmes; that is why the role of the suggested environmental advisory council should be strengthened. Consultation also implies a shared responsibility and accountability for all participants in the process.

Administrative Effectiveness and Efficiency

This can be attained by streamlining the administrative procedure established to enforce compliance of rules and regulations. It is also important to keep a constant vigil to remove ineffective or inefficient components of the administrative procedures.

Timely Decision-making

Nothing frustrates a dedicated environmentalist or a socially-conscious industrialist more than a time-consuming, rule-bound, and abuse-oriented bureaucracy. Consequently decision-points within the government environmental protection ministry must be clear and decisions made within a reasonable time frame.

Accountable and Responsible Environmental Management

The environment is being watched by everyone simply because it is the concern of everyone. Both the officers and senior management (including the minister) will have wide-ranging powers and authority, akin to police power. The temptation to abuse such power will therefore be great, especially in a developing country where land resources are limited and economic opportunities are many. Hence the officers will have to resist the forces of corruption, as much will depend on how the public perceives the credibility of the environmental ministry.

This statement, although directed towards public servants at large, is equally applicable to the officers of a department of the environment if they wish to be perceived by the public as doing their job responsibly, equitably, and accountably.

TRAINING OF ADMINISTRATORS FOR ESSD

There is a severe shortage of managerial personnel in all areas of governmental endeavour in Third World societies and the problem is particularly acute with respect to the task of environmentally sound and sustainable development (ESSD). Therefore, it is desirable that the developing nations draw up a blueprint for a national training strategy. This does not mean that other training programmes in place are to be overlooked or forgotten. On the contrary, such a strategy would integrate both aspects of training requirements.

It seems that in order to handle the administrative issues arising out of ESSD, a new paradigm ought to emerge as suggested below.[14]

Paradigm for ESSD Administration

In general terms, the suggested paradigm for ESSD could be configurated around the following traits:

(1) *Complexity*: administration for ESSD involves many and diverse elements. The importance of these as well as their interrelations are equally varied and diverse.

(2) *Contextualism*: administration for ESSD occurs in a multiple and complex context: economic, social, cultural, political, technological. This context both affects and conditions the institutional arrangements and the administrative system created for the purpose. Further, the context gives the administrative system its specificity and content.

(3) *Historicity*: the milieu in which the administration for ESSD takes place is not only structural; it is also historical. Existing institutions and the prevailing administrative culture and style conditions the process and procedures adopted for managing sustainable development. For example, the colonial legacy affects the present and future circumstances.

(4) *Relationship between short and long range framework*: administration for ESSD cannot be based on short-term thinking; protection of the commons and our natural legacy requires a long-range perspective. In the past, it is short-term gains which have affected our vision, and as a consequence we are now facing an environmental crisis of global proportions.

(5) *Relationship between micro and macro levels of administration for ESSD*: similarly, specific administrative practices and structures required for ESSD have to be seen in relation to broader perspectives and vision.

(6) *Heterogeneity*: the concept of sustainable development envisions interrelated and interlocking forms and manifestations of administrative requirements. Interconnectedness among the old and new institutions that have been created for environmental protection and conservation is a key to the success of ESSD.

Identifying Training Needs for ESSD Administrators

There is a need to make a concerted effort in Third World countries to provide standard education and training for ESSD administrators. In the past, training of public-service administrators has tended to concentrate upon generalists to take care of the more traditional type of executive law and order administration; this emphasis has to be changed so that training programmes are developed for specific developmental tasks. A training system for ESSD programme managers can be constructed in the same way as the more conventional scientific and technological systems in law, medicine or other traditional careers are organised, or it can be thoroughly innovative. For instance, it could integrate a series of levels in the system, from the scientific to that of the practitioner. An important aspect here is the need to recognise that managerial skills required for the tasks ahead in the 1990s and beyond are more complex. Further, management – as programme implementation and problem-solving – is, in the case of underdeveloped countries, most crucial at the grass-roots levels of the political process.

Public management involves essentially a public service. The essence of management here is not synonymous with business management. The training of managers for sustainable development cannot be just a transplantation of the business experience and methods, or the old-style public administration methodology, into the realm of present-day governmental process. All too often this distinction is ignored when it comes to providing the initial training for administrators. Yet the distinction between the provision of a service to a member of the public or a public good and the profitable bottom line remains fundamental. This distinction should be clearly ingrained in those who wish to enter the complex realm of development administration.

Towards a New Administrator for ESSD

This discussion has very direct implications for defining the tasks and the training programme of the new type of administrative cadre. For this, we may have to look at a new scientific and technological system to create the appropriate administrative know-how. If the administrators for ESSD are to be a new breed of managers, a number of formative conditions along the lines of the paradigm outlined earlier must be met. In operational terms, this entails the necessity to combine techniques with training in both the social and the natural sciences so that these administrators are ready to analyse and act upon the needs of sustainable development. In other words, an ESSD administrator also becomes a policy analyst with equal capability for operationalising and problem-solving; this may be possible only when such an administrator is equipped to appreciate the scientific and technical issues (including even rudimentary knowledge of environmental sciences and engineering), and can balance these by integrating them with policy sciences. Furthermore, given the globalisation of environmental and sustainable development issues, such an administrator would have to think in terms of global implications of his or her local issues. Thus, an ESSD administrator would need to acquire an international perspective and be capable of seeing global and regional trends which affect the process of development within his or her domestic circumstances. Acquiring such a perspective would be a prerequisite for a properly trained ESSD administrator.

NECESSARY INGREDIENTS FOR EFFECTIVE ENVIRONMENTAL ADMINISTRATION

The following are suggested as necessary ingredients for ensuring effective environmental protection administration strategies:

(1) The environment ministry of a nation should become a scientifically and technically driven organisation staffed by a multi-disciplinary team put together for proper planning, administration and enforcement of the national environmental legislation and programme. This requires a highly qualified staff, with talents appro-

priate to their function: broad training or professional experience, well developed analytical skills, and a superior interpersonal capability.

(2) In most Third World nations, officers of the newly established ministry of environment lack appropriate professional training and experience to handle the complicated task of implementing and enforcing the environmental law. They require adequate professional training and on-the-job experience. It is essential that all nations consider developing a training strategy for environmental protection and administration.

(3) Environmental regulation-making should be based upon consultation with industry. Evidence in some industrialised nations suggests that standards and criteria developed without reaching a consensus with industry result in a negative impact upon the industry being regulated. The regulatory process generally entails great discretionary powers in the hands of government authorities. Discretion implies flexibility; and yet the same discretion if it is poorly managed or unfairly applied may lead to arbitrariness or corruption.

(4) Whenever a new government ministry is established, there is a tendency for other ministries to accelerate the production of regulations and standards in order to establish their own jurisdictional pre-eminence, while at the same time the new ministry tries to secure its 'rightful' place in the system. Thus starts a regulatory war among these ministries which results in duplication and system-wide confusion. Only through a super policy co-ordinating body can jurisdictional peace be secured.

(5) Each ministry should develop an environmental assessment compliance monitoring policy/ programme so that its recommendations, conditions and terms of approval of EIA can be monitored. Further, the proponent should be required to submit an annual report on the undertaking for review by the ministry. In the field of environmental assessment, all relevant ministries and parastatal bodies should have the opportunity to review the environmental impact assessment submitted by a proponent and submit their comments or recommendations within an agreed time-frame.

(6) In several countries, there is a lack of the facilities necessary for testing samples taken from the sources of pollution. Even where such laboratory facilities exist there is little state-of-the-art equipment and trained personnel. While existing laboratories will in any

event have to be strengthened, there is a need to consider establishing a central environmental laboratory which could provide the necessary quality assurance and quality control (QA/QC) procedures and other testing requirements. Further, in order to avoid jurisdictional disputes among various labs, as well as in order to establish a proper mechanism of co-ordination, a memorandum of agreement concerning the responsibility, authority and jurisdiction of each of the labs should be created.

(7) It will be desirable for local authorities, specially municipalities, to consider entering into memorandums of agreement among themselves with respect to the construction, operation, maintenance costs, modification in design, the receipt and disposal of waste, a wastewater treatment plant, or the sewage works so that the present and future aspects of such undertakings can be firmly handled. Further, liaison committees should be established by each local authority to provide a forum for keeping the public informed of various projects, such as construction and operation of a landfill site, a sanitary sewage system, or a service for the recycling of waste.

(8) A joint government–industry research programme should be undertaken to assess the feasibility of adopting various pollution control technologies for the nation. Research and development programmes should be established early on, particularly as appropriate technology for industrial pollution control will be needed. Further, each country should consider investing in environmental biotechnology in co-operation with their national universities, industrial groups, and relevant government ministries.

(9) When there are several ministries involved in the wider aspect of environmental conservation and protection, sometimes other ministries feel as if they are at the receiving end of the directions emanating from the central environmental ministry. At the same time, they do not know if their plan of action or the methodology they may be using for enforcement (such as inspection, sample-gathering, getting the sample tested, documentation preparation, and presenting evidence before the court or an environmental tribunal) is compatible with other enforcing agencies. They may also wonder what will happen should two or more enforcing agencies become involved at the same time in the same situation. How will the confusion be resolved? Therefore, there is a need to develop a national policy and procedure for the enforcement and prosecuting of environmental crimes.

(10) Nowhere is the dictum *audi alteram partem* (the duty of a democratic government to provide the accused with a reasonable opportunity to present his or her case to an unbiased party for hearing)
more applicable than in the case of environmental offences. In
order to expedite the process of appeal from the industry or the
public arising out of the power of enforcement and regulation exercised by the ministry of environment, it is essential that an independent, quasi-judicial tribunal or appeal board be established.
Such a tribunal would act as an arbiter of complaints from the public and industry, as a complement to the existing judicial process of
appeal.

A THIRD WORLD PERSPECTIVE

Interaction between the rich North and poor South on these issues will
have to be based on a certain accommodation. From the Third World
perspective, the following items appear to be a part of the global
demands which they would like the industrialised nations to consider for
the protection of the planetary environment:

1. *Internationalisation of the 'Polluter Pays' Concept*

There is a need to internationalise the Polluter Pays' Principle. At
present, the concept is being used by some industrialised nations, but
only within their own borders. However, they do not recognise that they
(or their multinationals) are the main polluters of our world and ought to
'pay' for it. Towards this end, the industrialised nations should establish,
based on their past and present contribution to the world pollution
aggregate, a permanent fund for cleaning up after their collective and
individual actions. Further, environmental technology transfer should be
available to all nations, free from any restriction such as copyright,
patent, and trademark.

2. *Green Imperialism*

Several developing nations have a basic distrust of the promises and
policies of the North. They think that the current emphasis on environmental problems by the industrialised nations is merely a continuation

of their economic domination and their trade barriers. Suggested constraints on development, barriers on their exports, and the requirements of structural adjustments programmes have led the poor nations to oppose strenuously any environmental conditionality. Further, the industrialised nations should take the lead in reducing per capita energy consumption, because over-consumption causes fast depletion of non-renewable energy resources, and at the same time affects the global environment. The North should also control automobile emissions at home and require their multinational motor corporations to introduce catalytic converters and the use of lead-free petrol.

3. *Consumption and Materialism, not Population, are the Culprits*

The South believes that the consumption patterns of the North and its continuous emphasis on a materialistic way of life are the basic culprits of global environmental problems, rather than the rising population among poor nations. For example, seen from the perspective of greenhouse gas emissions, the United States emits 1000 billion tons of carbon gases while India's share is only 230 billion tons, even though the latter's population is four times greater. Or consider other indicators: in 1990, 251 million US people used more energy for cooling their homes than 1.1 billion Chinese used for all purposes; the North accounts for 73 per cent of total industrial carbon dioxide emission in the world. Production of CFCs should be halted; alternative technology for refrigeration and air-conditioning must be found and made available to the Third World nations.

4. *Differing Perceptions about the Global Environmental Problems*

North and South differ about the priority to be given to various environmental issues. From the South's viewpoint, poverty alleviation is the most crucial task of the day, and not ozone depletion or release of hydrocarbons and the like. As Mrs Indira Gandhi stated during the 1972 UN Conference on Human Environment in Stockholm:

> On the one hand the rich look askance at our continuing poverty – on the other, they warn us against their own methods. We do not wish to impoverish the environment any further and yet we cannot for a moment forget the grim poverty of large numbers of people. Are not poverty and need the greatest polluters?[15]

It is obvious that not much has changed during the past twenty years since Mrs Gandhi made that statement. Developing nations are still struggling with poverty and hunger, and attempting to provide basic health care to their people. To most of them, environmental concerns are the affliction of over-development.

5. The Need for an International Environmental Monitoring/ Enforcement Agency

At the UN level, an organization should be set up to monitor the state of the environment of our planet, to prepare an annual environmental audit which would report on enforcement of international environmental treaties and conventions, and on infractions caused by individual nations as well as by multinational corporations. If necessary such cases should be brought before the International Court of Justice. No country would have veto power over this organisation.

The Third World countries cannot afford the costly react-and-cure approach to dealing with environmental problems – problems that are becoming increasingly more complex and pervasive and more difficult and expensive to clean up. Further, these nations can no longer afford to ignore environmental considerations in properly assessing economic performance. Environmental degradation and mismanagement of their environmental resources impose social and economic costs.

At the same time, another thing should also be made clear. The people of the Third World are going to face some difficult choices. People are likely to see the costs of environmental protection reflected in higher prices. But these costs will be offset by improvements in the quality of the air and water, and ultimately, in human health. Moreover, new economic opportunities will be provided by a growing environmental industry sector – opportunities that governments and businesses must explore, given the challenges of an increasingly integrated and more competitive world economy. In the final analysis, the economy will experience net gains as a result of integrating environmental consideration into decision-making.

The World Development Report, issued by the World Bank in May 1992, recommends that the high-income countries assume the primary responsibility for addressing the world-wide environmental problems – green house warming and depletion of stratospheric ozone – of which they are the primary cause. Further, as the rich nations receive major benefits from the protection of tropical forests and bio-diversity, they should bear a proportional cost of the protection efforts undertaken by the Third World nations. The report recommends that greater access be given to the poor nations to 'less polluting' technologies; it also suggests that priority be given to bringing safe drinking water and adequate sanitation to the approximately one-third of a billion of the world's population which does not have them.

The report argues that the key to sustainable growth is not to produce less, but to diversify production. At the same time, institutions for environmental protection must be strengthened and current agricultural and industrial policies changed so as to reduce drastically the amount of pollution, wastes, and other environmental damage per unit of output. Further, in those cases where problems require action which goes beyond the capacity or self-interest of a nation, the industrial countries should provide funding to such world-wide problems. Finally, the report notes that alleviating poverty is a moral imperative and a prerequisite for environmental sustainability. This is the greatest challenge for the world.[16]

CONCLUSION

The Transfer of Pollution-Free Technology[17]

Although developing nations would gladly accept changes in technology, they are worried that they may be given technologies by the North which were being used there about two or three decades earlier, the very methods that brought us to the brink of environmental disaster we face today. Developing nations require technology to be transferred for four major purposes: for cleaner and more efficient production; to minimise energy requirements, industrial waste, and pollution; to prevent air and water pollution; and to get agreements on certain international conventions such as the Montreal Protocol so as to mitigate the adverse impacts of environmental damage caused, in the first place, by the industrialised world.

Another consideration which worries the South is the issue of intellectual property rights and commercial profits by the multinational corporations. In this respect, a new global partnership is required to create a mechanism to make financial resources available at moderate and reasonable rates, and to transfer environmentally sound technology on non-commercial terms so that the developing nations could make a swift technological transition. The international community must appreciate that the current culture of the West – based on mass production, mass consumption, and mass waste generation and disposal – would have to change too. A commitment from the North would have to come on two fronts: to change its life-style, and to provide assistance to the developing nations for environmental conservation and protection, not in the form of bilateral aid or multilateral loans but as partnership-grants between nations.

The Globalisation of Environmental Issues

People living in all parts of the world can be held together in this common enterprise of protecting our common heritage by the values and ideals that each community subscribes to, as well as by self-interest in the best sense. By combining their concerns, talents, energies and resources, people can build a common future on earth in which they can share their fortune (meagre or in abundance) to equalise opportunities among all their neighbours and among all the regions, and to sustain the right of all and everyone to fulfil their ends in life. That is why the globalisation of environmental issues is important.

One of the means to encourage co-operation among people of the world is to foster the convergence of the shared fundamental value: protection and enhancement of the environment. Of course, there are differences in life styles across the globe, and people do not want identical lives, cultures and beliefs. Nevertheless, there are several mutually reinforcing rather than incompatible cultural values which provide us with a remarkable degree of agreement. Given the opportunity to express and act locally on these issues, communities can respond to regional and global challenges. For this, reforms will be needed in international laws and treaties, and people's attitudes will have to go through some fundamental changes with respect to their life style, and an ability to reconcile the needs of different communities and local interests will have to be carefully nurtured.

The reality of our world – the central fact that we must take into account – in the design of global environmental policy and programmes is not independence but interdependence. The complexity and scale of environmental problems no longer permit a water-tight division of environmental issues. The challenge before us is in the reform and renewal of the existing world system, and in securing a firm place for the rights of future generations, by bringing into reality an environmentally sound and sustainable development. It is to this end that much of our effort must now turn, during these last years of this twentieth century, if this planet of ours is not to be further damaged. This challenge is not limited to the Third World alone; we are all in it together.

Notes

1. M.S. Swaminathan, 'The Environment Scene: Maladies and Remedies', *The Hindu: Survey of the Environment 1991*, Madras, April 1991, p. 5.
2. World Commission on Environment and Development, *Our Common Future* (New York: Oxford University Press, 1987) p. 8.
3. World Bank, *World Development Report 1992: Development and the Environment* (New York: Oxford University Press, 1992) p. 34.
4. Ibid., p. 8.
5. World Commission on Environment and Development, p. 310.
6. Ibid., p. 312.
7. This is based on a report prepared by Catherine Eiden, Environmental Science Adviser to the Ministry of Environment and Quality of Life, Government of Mauritius, Port Louis, 30 April, 1991.
8. This framework is based on the author's report prepared for the Ministry of Environment and Quality of Life, Government of Mauritius, and published as *State of the Environment in Mauritius*, edited by O.P. Dwivedi and V. Venkatasamy (Port Louis: Government of Mauritius, 1991) pp. 12–15 and 295–302.
9. This description is based on a report prepared by the author for the Government of Mauritius, *White Paper on National Environmental Policy for Mauritius* (Port Louis: Ministry of Environment and Quality of Life, 1990).
10. This perspective was developed in a report by Frank Work, legal adviser to the Government of Mauritius, in July 1990.
11. For further elaboration, see *State of the Environment in Mauritius*, pp. 269–72.
12. The author developed this approach as part of a report published by the Government of Mauritius; see *White Paper on National Environmental Policy*, p. 2.
13. See World Bank, *Economic Development with Environmental Management Strategies for Mauritius*, Report No. 7264–MAS (1 November 1988) pp. 2, 12.
14. This is based on the reformulation of a similar paradigm prepared by the author, in association with J. Nef and published as an article, 'Training for

Development Management: Reflections on Social Know-how as a Scientific and Technological System', *Public Administration and Development*, Vol. 5, No. 3, 1985, pp. 237–48.

15.   Quoted in O.P. Dwivedi, 'India: Pollution Control Policy and Programmes', *International Review of Administrative Sciences*, Vol. 43, No. 2, 1977, p. 123.

16.   *World Development Report, 1992*, pp. 1–24.

17.   For further elaboration, see O.P. Dwivedi (ed.), *Perspectives on Technology and Development* (New Delhi: Gitanjali Publishing House, 1987).

6 When Means and Ends are at Variance: Administration for Sustainable Development

INTERNATIONAL INDUCEMENTS FOR ADMINISTRATIVE REFORMS

Chapters 2 and 3 have demonstrated that administrative reforms suggested to the Third World countries were derived from changes being made in the internal administrative systems of the industrialised nations themselves. Since the 1950s a parade of administrative reforms has been suggested, starting with that of Paul H. Appleby from the United States. His report, *Public Administration in India*, prepared for the Government of India in 1953, very clearly suggested that India should consider strengthening 'democratic administration':

> Democracy hinges first of all on the manner in which responsibility is fixed and held accountable; second on responsiveness and considerateness. There are techniques that enhance responsibility and accountability, that enrich responsiveness and considerateness; these are democratic techniques. There are methods that diffuse and conceal responsibility, that reduce accountability, that misinterpret responsiveness, that over-burden citizens and that convert considerateness into sticky sentimentality. These damage effectiveness and demean democracy.[1]

Of course, the techniques Appleby had in mind were based on what was happening or ought to have happened in the United States. Appleby, to remind ourselves, was not the first foreign expert to suggest reforms in the administrative practices in a former colony: the first such report was prepared by Lord Macaulay for the East India Company in 1854, which recommended the 'rank-oriented concept' for civil service appointments and promotions. This was later used by the Northcote–Trevelyan Report, which became the foundation of the modern British civil service system. The Macaulay Report was considered so crucial that it was reprinted by the Fulton Committee which submitted its report on the British civil service in 1968.[2] But India was not the only developing country where external experts were sent in to advise the government. No developing country was spared this experience, and sometimes experts from different western nations suggested contradictory reforms.

Towards the end of the 1980s another important factor broadening the scope and spectrum of Third World problems was the withering away of Eastern European communist states and the entry of at least some of them into the realm of the Third World. While the eastern bloc nations may, in general, be considered 'poor' for the time being, many of them may not remain in this state long. Akin to the economic recovery of Western Europe during the post-war era, these nations are going to leave the Third World nations behind.

If the downsizing of bureaucracy continues, by the turn of this century there should be fewer public service officials as a proportion of the population, even though some systems will continue to expand. Certainly, the rate of growth of public sector expenditures and personnel may be reduced as the emphasis on privatisation grows. Public servants at all levels and in all areas of the world should be better trained, more professionally oriented, more aware of the world at large rather than concentrating on their confined environment, more ethical, more productive, and perhaps more humble. If this happens, morale will improve; the public's attitude towards public servants will change for the better; and development administration will co-exist happily alongside private sector entrepreneurship, as public officials provide a positive, enabling environment for economic expansion, unencumbered by red tape.

Since the mid-1980s, the rush to overhaul the administrative systems of the Third World has been prompted largely by the conservative policies pursued by the governments in the United Kingdom, Germany, the United States and Canada. The overhaul, it seems, is not towards increasing the effectiveness and reach of governments but rather is based on the premise 'how to get more for less'. The concept of a new 'managerialism' has been floated – a concept that means pruning staff, advocating cost consciousness, the adoption of performance measurement, non-scale remuneration, and the creation of non-ministerial agencies beyond the jurisdiction of the public service commission. Suggested remedies such as business methods, value-for money and the Three Es (economy, efficiency, effectiveness) are all based on the conservative perspective of introducing business methods into the public service. As noted by Andrew Dunsire, it seems that 'the whole edifice of restructuring of 'public administration reform' – liberal, divestment, consumerism, as well as managerialism – balances on this single assertion, that public servants ... are not to be trusted with the public interest.'[3]

It would appear that a new orthodoxy is emerging which has the following elements: reduction in the role of government, market-friendly economy, deregulation, divestment, consumerism, and use of business methods in public administration. This neo-orthodoxy is being espoused by the World Bank, the IMF and the major donor nations such as the United Kingdom, the United States, Canada, Japan and Germany.[4] This has followed on the heels of high rates of inflation, frequent currency devaluations, downsizing the public bureaucracies, privatisation, value for money, unprecedented levels of servicing of foreign indebtedness, and constant pressures from the World Bank and IMF to restructure the economic and administrative systems. But the structural adjustment programmes (SAPs) pursued by the World Bank and IMF, with experience of over a decade in Asia, Africa and Latin America, are still viewed with scepticism by many. People have seen that SAPs, with the exception of a very few cases, have brought high human costs, political instability and economic disasters. Yet despite mixed results, it seems that the international pressure for SAPs will continue.

In spite of the West's enthusiasm, certain myths about the new orthodoxy have been exposed: governments cannot be run like a private industry; administrative excellence cannot be transplanted from foreign administrative cultures or from their manuals of management; politicians are unwilling to reduce interference in and control of governmental bureaucracy; and privatisation/divestment has generally meant that the few well-run profit-making public enterprises are handed over to a few wealthy people while there are no takers for the many loss-making, inefficiently run companies. As noted by the *Guardian*:

> Nearly all of the recently privatised organizations were profit making enterprises, geese that regularly and reliably laid golden eggs which went into the public purse to benefit the entire population. The government then under-valued them and sold them. Who to? Why, the richest twenty percent of the population. Now the geese are still

laying golden eggs, but for the benefit of only twenty percent of the population, if that.[5]

In the context of a developing society, the basic question remains unanswered: when one transfers ownership from the entire population to a small, wealthy private sector, does it not really mean stealing a national asset? It should be noted that there is no such thing as a universal remedy for all ills, be they economic, social or administrative. Changes happening in the West, such as reconstruction, public administration reform and 'rolling back the state' may be more relevant there, but they are not necessarily a cure for the public administration problems of a developing nation.

In its *World Development Report for 1991*, the World Bank stated that the role of government is larger than merely controlling the economy; instead, governments should invest in education, health, nutrition, family planning and poverty alleviation; building social, physical, administrative, regulatory and legal infrastructures of better quality; and mobilising resources to finance public expenditure. Reforming institutions is seen as the key to the very core of development. 'Reform of the public sector is a priority in many countries. That includes civil service reform, rationalizing public expenditures, reforming state-owned enterprises, and privatization.'[6] Strengthening of institutions and investment in human development have been key elements in the World Bank's current mission.

Of course, administrative reforms are necessary. However, before a Third World country accepts blindly the suggested changes from external sources, it should consider whether such changes are going to be beneficial in the long run, and whether they will outweigh the costs. Any organisational change adds complexity and disruption to the existing equation and requires adjustments in the administrative culture of the nation. And while the change may be tempting, and perhaps a way out – by apportioning the blame to the previous regime – the nation may suffer in the long run if the risks, costs, and transferability are not appropriate. In order to bring such a change, a burning spirit of righteous indignation is desirable so that bureaucratic inertia, political indifference and public apathy can be overcome. As Gerald E. Caiden pointed out, 'Reform can only control, not cure, the ills of public administration. It exposes maladministration and proposes alternatives that promise to reduce it but it cannot guarantee good administration nor prevent the reappearance of maladministration.'[7]

ORGANIZING TO GOVERN

Administration for sustainable development will depend on the difficult task of organizing to govern. It should be noted that the structure of government as it exists in all the developing nations did not evolve according to any master plan, although various efforts were made to plan activities relating to the governance of a nation. Governing processes have evolved incrementally, mostly in response to political concerns, the personal visions of some leaders (mostly at the time of independence), regional interests and ethnic and tribal loyalties. But most important has been the influence exerted by the industrialised nations and by the World Bank and IMF.

There are several perspectives to consider here. Sometimes the governing process suffered from too many goals proposed by various international organizations, thereby creating confusion and organisational constipation. Then there came a time when what the government wanted to accomplish, what it was being forced to do, and what resources it had to pursue such objectives, all became blurred. What was not realised was the fact that a change in any building block of governance would have serious implications for others, if not for the entire system. The second problem came when the international organisations, as well as international aid agencies from individual countries, did not appreciate the timing of their recommendations for effective changes. It is well known that the greatest opportunity for an organisational overhaul is immediately after a general election or even after a *coup d'état*. That is the time when public expectation that the new government will bring about the needed change is high. It is a time when the new government has the maximum degree of freedom to set the direction it wishes to take so as to fulfil its mandate. However, the opportunity to make changes wanes while international organisations such as the World Bank and the IMF, and the international aid agencies, take time to think of what to do. Finally, governing in the Third World has become as complex as in the North: in some countries a new government will appoint a great number of persons to the cabinet whether or not such a number is warranted. A large cabinet of ministers not only makes the governing process cumbersome; it also leads to costly empire-building, is time-consuming, and is difficult to co-ordinate. In order to focus clearly on its goals, the size of the government team ought to be manageable.

SPECIFIC ISSUES

Some specific issues are discussed below which relate to the focus of this chapter and monograph:

Self-reliance with Minimal External Help

The poor are almost completely self-reliant, and they always use whatever resources (including networking) they can muster among themselves. Similarly, the poor countries have to think in terms of how best they can use their indigenous know-how. They should treat external aid as non-existent, and instead turn to self-help projects based on the needs of their communities. Large-scale projects have often excluded the great majority of the poor; examples abound concerning the displacement of people due, for instance, to large-scale hydro-electric projects in many developing nations. It is generally true that it is difficult for the very big to notice the very small. Sustainable development will have to begin with the poor of the Third World. Perhaps here the following words of Jules Feiffer are relevant:

> I used to think I was poor. Then they told me I was needy. Then they said it was self-defeating to think I was needy; instead I was deprived. Then they said deprived had a bad image; I was really underprivileged. Then they said underprivileged was overused; I was disadvantaged. I still don't have a cent, but I have a great vocabulary.[8]

The Globalisation of the Theory and Practice of Administration for ESSD

Keith Henderson has pleaded for the need to recognise the trend towards the globalisation of administration for development.[9] Nowhere is this more applicable than in the sphere of environmentally sound and sustainable development. This constitutes a process by which administrators and scholars can participate in, as well as contribute to, the development of the worldwide theory and practice of SD administration. Because it involves a common core of knowledge about development which ought to be environmentally sound and sustainable, and as this knowledge may be available for any part of the world, it is desirable that open-ended thinking prevail so that the policy-makers in developing nations know more about the world at large than what lies within their national boundaries. This is a kind of 'collective self-reliance'.[10] It is here where there is a place for convergence of indigenous and international thought and practice. The environmental threat facing our planet earth has brought us all together. This calls for an increased amount of interdependence and connectedness in a complex yet shrinking world where all of us have to survive.

The End of The Old Order

Starting in 1989, the world witnessed the collapse of several authoritarian regimes, including the bulwark of communism in the former Soviet Union and Eastern Europe. Now, in many of these countries, a multiparty democratic system has been introduced, while in nations such as Cambodia, Ethiopia, Angola, Nicaragua and Peru, leftist governments have been replaced or are in the process of getting changed. All over the world, the old order, which came into being after the Second World War, is disappearing. Democracy, the market economy and human rights are the new slogans.

However, the ushering-in of a new era has not produced peace and economic progress. History is being repeated, in the sense that when countries such as India, Nigeria, Ghana and Indonesia received their independence, there was a huge expectation that freedom would bring peace and economic progress. This did not happen, despite the promises made by some political leaders to their people that once independence was gained, no one would remain poor; on the contrary, poverty, ethnic rivalries and religious wars were visited upon them. This is what is happening in Yugoslavia, and to a lesser extent in Georgia, Armenia, Azerbaijan, Ethiopia, and in various other African countries. Age-old animosities between various nationalities have cropped up under the rubric of self-determination.

It seems that sharper conflicts are taking place in the name of democracy, human rights, religious freedom and the right to self-determination, and the sense of accommodation and the spirit of compromise which kept various nations together is disappearing. It is not yet clear whether more balkanisation will take place, and whether this will result in further cleavages, religious intolerance and wars. Can the western ideals of democracy and human rights work in those states which suffer from chronic poverty and maladministration? Can the new order guarantee peace and economic prosperity if the various conditionalities are not met?

Incomplete Development

In the past, some damage to the environment was accepted as a necessary cost of economic development; this attitude is no longer justified. The style of development imposed on Third World nations has been only partly successful; in the main it has been patchy and incomplete. Developmental planning and aid has not been able even to reduce poverty, and with increased poverty has come a more deadly impact on the environment. One unwanted side effect of incomplete development has been the growth in population size, because the mortality rate has fallen and life expectancy has increased but with little positive impact on birth rates. It has been demonstrated that improved family income and welfare reduces the birth rate. Thus, the removal of poverty is crucial for controlling population growth. Another side effect of incomplete development is the massive growth of Third World cities. 'Urban growth has spiralled beyond control in many developing countries, and is causing not only an urban but a rural crisis. Together they add up to an environmental and developmental disaster.' [11]

These two side-effects will not disappear as long as there is a perception that development in Third World nations can happen simply by replicating the experience of the West. There is a need for a radical change in approach to development. From the perspective of this book, one such approach is the adoption of a strategy towards environmentally sound and sustainable development.

Balanced Institutional Development

During the first four development decades, the developing countries adopted interventionist and regulatory policies for their societies. At the time, central planning and state control over the economy were deemed to be crucial, which resulted in the nationalisation of various commercial, business and banking interests; it also created a huge number of public enterprises. But by the 1980s it became clear that these policies were counter-productive when it was found that most government-owned and administered enterprises were inefficient, uneconomical, overly bureaucratic and corrupt. Excessive regulation and intervention in the economy breeds corruption; this is evident in many developing nations.

The Cultural Divide and Enrichment Across Cultures

It is generally believed in the West that principles and concepts are universally applicable, and are transferable (with some modifications) to other nations or organisations.[12] This belief is based on the premise that, as scientific theories have universal application all over the world, the same notion ought to be applicable to other human endeavours. Perhaps this view would have remained unchallenged had it not been for the Japanese (and later South Koreans, Taiwanese, Hong Kong residents and Singaporeans), who proved otherwise by their economic miracle and unique management style. As we know now, by the 1970s the Japanese had surpassed even the United States' productivity. The Japanese way became an alternative management technology for doing things. Later, when South Korea, Taiwan, Hong Kong and Singapore joined the Japanese in this process, people in the West had to acknowledge that simply because a 'principle/concept' has originated in the West, it does not automatically become the only truth.

Although leaders like Mahatma Gandhi had expressed a similar view, it was not until the 1970s that an appreciation of the cultural divide became accepted. It became evident that the style of doing things does differ from culture to culture. For example, US society is individualistic, where each individual is the most important unit; hence efforts are made to let that individual fulfil his or her own potential. On the other hand, in Japan society tends to care for groups: people identify themselves with the company or the organization they belong to, and it is within this group that the individual finds his or her identity and purpose in life. Similarly, in other parts of Asia and Africa one finds the continuing influence of tribal, caste, sect and religious affiliations to be very strong.

This raises three specific questions. If the concepts and principles of doing things (that is, the management style) differs across cultures, why should there not be a different developmental paradigm for the Third World? If the root of such differences between the West and the South is culture, then should not 'culture' be the foundation upon which one should build alternate models of development? And finally, how long will Euro-centrism or North-centrism continue to force its own cultural paradigm on to others with its push for a single-market world, profitable to the North-dominated development industry? These questions are being raised not to refuse the technical assistance of the West nor to reject its scientific/technological ascendency, but rather to question the transferability of western concepts *in toto* to a culturally alien society. What is being suggested is that the 'power to shape ideas and events', which has so far lain with the West, needs to be shared now. This is nowhere more evident than in the field of environmental management.

These principles can be summarized in the form of some generalised statements.

Mutual Learning

Transfer of an institution to another cultural or geographical setting should be beneficial not only to the recipient nation but also to the donor. For example, the Canadian system of environmental adjudication was proposed by the author when he was environmental policy and institutions adviser for Mauritius in 1991. But the cultural and administrative context of its application were not taken into consideration, with the result that later in 1993 the government of Mauritius decided to change this provision in its Environmental Protection Act. To consider another example, when the Bhopal tragedy struck India in December 1984, killing about four thousand people, Environment Canada immediately appointed a task force to study the catastrophe, and to learn from it so that an appropriate response and preparedness could be instituted in Canada.[13] Moreover, when an institution is exposed to an alien environment, it gets tested thoroughly. That is why both nations (or North as well as South) become beneficiaries in the transfer of knowledge.

Creative Tension Between theViews of North and South

As mentioned in Chapter 5, there is an imbalance between North and South with respect to intellectual technology. What is needed now is to restructure this inequality. For this, the decolonisation of the mind becomes a precondition for intellectual maturity and freedom of vision. This does not mean that concepts and technologies of the North should be avoided; otherwise, the South may retreat to reactive nativism.[14] Further, there is a need to develop a genuine dialogue and a fair exchange of ideas between the two groups. Out of this interplay, there may emerge new ideas which could be equally applicable in North or South, instead of the one-way process which has been the case so far.

The Need for Comparative and Cross-Cultural Learning and Research

When students from the developing nations arrive for higher studies in the universities of North America and Western Europe, they are forced to blend in with the existing curriculum which is best-suited to the local society. Thus, when they complete their education they become skilled with the analytical tools and methodologies of the North. To give an example: several years ago, I was baffled to note in a Senate meeting of the University of Guelph that a doctoral student from Ghana did his research on winter wheat. When asked a question concerning the relevance of this four-year research for the student's society, the supervisor advised that although the research in question had nothing to do with agricultural production in Ghana, the funding under which this student's research was done in Canada came from the Ontario government for winter wheat research. Thus, in general, the domestic concerns of Canada and other countries of the North control the outcome of research irrespective of the cultural background of students. On the other hand, if the faculty and student could have been exposed to a comparative and cross-cultural aspect of learning and research, the result would have been mutually beneficial.

Protecting Against the Rapid Loss of Traditional Cultural Legacies

The overwhelming impact of western culture, which gets transmitted generally in its rudimentary and consumption-oriented behaviour, has created a rapid loss of traditional cultural legacies in the South. An added dimension to this aspect is the break-up of traditional units of community, such as village and extended family systems. And then discontinuities are being created by the young because they consider such traditional cultural norms irrelevant seeing them as anti-modern and backward. Their mythical vision of the North is creating a gulf between the present rulers and successive generations; they do not realize that the bastardisation of traditional culture would make their generation a mixed and confused group. Perhaps integrating the old with the new (from the West) could serve the purpose. Nevertheless, without protecting the rapid loss of authentic traditional cultures, symbols, myths, and ethos, their entire society and polity will continue to be impoverished.

There is no doubt that each society has to consider what is good for it in the long run. Governments in the Third World should raise serious questions about the cultural and social relevance of the transfer of Western paradigms and technology; resist the temptation to be swayed by the emerging fads in the West; reflect back on the cultural relevance of the recommendations made by international aid agencies; and try to evolve their own indigenous brand of concepts, design and intellectual technology to deal with distinctive domestic and external issues. The key to success will greatly depend on how these concerns are integrated in a systematic way to evolve a comprehensive vision for the welfare of all of humanity.

ADMINISTRATION TO MANAGE COMMON INTERESTS AND THE COMMON GOOD

Ends and means at the disposal of Third World nations have always been at variance because the centre – that is, the West – has maintained control; however, this situation cannot continue any longer, as demonstrated by the Earth Summit in June 1992. If people living in all parts of the world are to be held together in this common enterprise of protecting our common heritage, they must come together and bring those values and ideals that each community subscribes to. By combining their concerns, talents, energies and resources, people can build a common future on earth. They will be able to share their fortune (however meagre or abundant) to equalize opportunities among all neighbours and regions, and to sustain the right of all to fulfil their ends in life. That is why the globalisation of environmental issues is important.

In 1945, when world leaders met and established the UN, the dominant thinking was of peace and security. Since then, the Cold War has ended, the possibility of nuclear annihilation has largely disappeared, and the world seems to be more secure, even though there are many local conflicts going on. When the 117 world Heads of State met at the Earth Summit in Brazil in June 1992, they met not to worry about the next world war but because the fear of another world-wide calamity was facing our planet. They agreed on sustainable development as their new mission; perhaps it may become the new rallying point for positive co-operation between North and South. An enormous task awaits the international community in the final decade of the twentieth century: that is, in the words of the Brundtland Commission, of meeting the needs of the present without compromising the ability of future generations.

One of the means of encouraging co-operation among people of the world is to foster the convergence of a shared fundamental value: protection and enhancement of the environment. Of course, there are differences concerning lifestyles across the globe; further, it is agreed that people do not want identical lives, cultures and beliefs. Given the opportunity to express and act locally on these global issues, communities can respond to regional and global challenges. For this to happen, some institutional changes will have to be undertaken: public service officers will have to be specially trained to consider the impact of their administrative decisions on the regenerative capacity of their natural and human resources; the compartmentalised thinking of governmental bureaucracies will have to be overcome so that all could work together towards the goal of human development; international laws and treaties must be reformed and the attitudes of people will have to go through some fundamental shifts with respect to their life style. In this process, mutual respect, flexibility, adaptiveness, and the ability to reconcile the needs of different communities as well as of local interests would have to be carefully nurtured.

Development administration in the past has been concerned mainly with transplanting and replicating ideas and institutions of the North. It continues to practice statecraft in a hierarchical, bureaucratic, and centrally planned manner. But as the events of 1989 have demonstrated, that strategy was one of the causes of stagnation in the former communist regimes of Eastern Europe. Developing nations must now employ alternative institutional designs to escape from further damage to their system. Slowly, a realism is setting in among the Third World leaders that changes (both in the developmental process as well as in the administrative system) are incremental, and that it will take time for statecraft to transform itself and to align with its own indigenous social and cultural milieu. Development administration of the future will have to be an integral part of their indigenous social and cultural systems; because unless it brings in behavioural change among those who govern, the opportunities for action will be lost and uncertainties will continue to plague the prospects for sustainable development.

Development, after going through various trials and tribulations during the past forty years, has emerged as a fragile and multidimensional process. Contraints to development are better understood today than before. For development administration does not any longer connote those overly optimistic visions and grandiose plans by political leaders of the South who promised heaven on earth, not realizing that there existed a built-in variance betweeen those ends and means at their disposal. But despite some regression, failures and disappointments of the past decades, there is an optimism in the air. The greatest need of the time is to instrumentalize new ideas into action. It is to this end that efforts must now turn, and it is here where development administration can help the people of the Third World to achieve their dream.

Notes

1. Paul H. Appleby, *Public Administration in India – Report of a Survey* (New Delhi: Government of India Cabinet Secretariat, 1953) p. 68.

2. See United Kingdom, *Report of the Fulton Committee: The Civil Service, 1966–68*, Vol. 1 (London: HMSO, June 1968). Regarding the rank-oriented classification system, see also J.E. Hodgetts and O.P. Dwivedi, *Provincial Governments as Employers: A Survey of Public Personnel Administration in Canada's Provinces* (Montreal: McGill–Queen's University Press, 1974).

3. Andrew Dunsire, 'The Politics of Public Administration Reform', *The Changing Role of Government: Management of Social and Economic Activities*, Proceedings of a Commonwealth Roundtable held in London, 4–7 June 1991 (London: Commonwealth Secretariat, 1991) p. 201.

4. Nasir Islam, 'Managing the Public Service under Structural Adjustment: Politics and Implementation of Administrative Reform', *The Changing Role of Government*, p. 215.

5. *Guardian*, 23 September 1989, quoted by Andrew Dunsire in *The Changing Role of Government*, pp. 195–6.

6. World Bank, *World Development Report 1991: The Challenge of Development* (New York: Oxford University Press, 1991) p. 9–10.

7. Gerald E. Caiden, 'The Vitality of Administrative Reform', *International Review of Administrative Sciences*, Vol. 54, No. 3, September 1988, p. 354.

8. Quoted in Nafis Sadik, 'Rethinking Modernism: Towards Human-Centred Development', in *Development* (*Journal of the Society for International Development*) 2, 1991, p. 19.

9. Keith M. Henderson, 'Rethinking the Comparative Experience: Indigenization versus Globalization', in O.P. Dwivedi and K. M. Henderson (eds),

Public Administration in World Perspective (Ames, Iowa: Iowa State University Press, 1990) pp. 333–41.

10. Gerald and Naomi Caiden, ' Towards the Future of Comparative Public Administration', in *Public Administration in World Perspective*, pp. 363–97.

11. Sadik, 'Rethinking Modernism', p. 16.

12. For further information, see Gabino A. Mendoza, 'The Transferability of Western Management Concepts and Programs, An Asian Perspective', *Education and Training for Public Sector Management in Developing Countries* (New York: Rockefeller Foundation, 1977) pp. 61–71.

13. Canada, *Bhopal Aftermath Review: An Assessment of the Canadian Situation*, Report of a Steering Committee for the Minister of the Environment (Ottawa: Supply and Services Canada, 1986).

14. See M.A. Qadeer, 'External Precepts and Internal Views: Dialectic of Reciprocal Learning in Third World Urban Planning', in B. Sanyal (ed.), *Breaking the Boundaries* (New York: Plenum Press, 1991) p. 204.

Select Bibliography

AGARWAL, SHRIMAN N., *The Gandhian Plan of Economic Development* (Bombay: Padma Publications, 1944).

APPELBAUM, RICHARD, *Theories of Social Change* (Chicago: Markham Publishing, 1971).

ARENDT, HANNAH, *On Violence* (New York: Harcourt, Brace and Jovanovich, 1969).

ARNOLD, H.M, 'Africa and the New International Economic Order', *Third World Quarterly*, vol. 2, no. 2, 1980, pp. 295–304.

BLAIR, P., *Development in the People's Republic of China: A Selected Bibliography* (Overseas Development Council, 1976) Occasional Paper No. 8.

BODENHEIMER, S., 'The Ideology of Developmentalism: American Political Sciences Paradigm-Surrogate for Latin American studies', *Berkeley Journal of Sociology*, vol. 15, 1970, pp. 95–137.

BOWLER, PETER, 'Will Science and Technology bring Conflict within Third World Cultures?', *Science Forum*, vol. 10, no. 3, June 1977, pp. 12–13.

BOWMAN, JAMES S., 'The Management of Ethics', *Public Personnel Management*, vol. 10, no. 1, April 1981, pp. 59–66.

BRAIBANTI, RALPH, 'Transnational Inducement of Administrative Reform: A Survey of Scope and Critiques of Issues', in John D. Montgomery and W.J. Siffin (eds), *Approaches to Development: Politics, Administration and Change* (New York: McGraw-Hill, 1966) pp. 133–83.

BRANDT, WILLY (Chairman), Independent Commission on International Development Issues Report, *North–South: A Program for Survival* (Cambridge: MIT Press, 1980).

CAIDEN, G.E., *The Dynamics of Public Administration: Guidelines to Current Transformations in Theory and Practice* (New York: Holt, Rinehart and Winston, 1971).

CAIDEN, G.E., 'Ethics in the Public Service: Codification Misses the Real Target', *Public Personnel Management*, vol. 10, no.1, 1981, pp. 146–52.

CAIDEN, G.E., 'The Vitality of Administrative Reform', *International Review of Administrative Sciences*, vol. 54, no. 3, September 1988, pp. 331–357.

CEBOTAREV, E.A., 'The Diffusion of Technology: Blessing or Curse for Latin American Nations?' in J. Nef (ed.), *Canada And The Latin American Challenge* (Guelph: Ontario Cooperative Programme on Latin American and Caribbean Studies, 1978) pp. 93–103.

CHATURVEDI, T.N., *Transfer of Technology Among Developing Countries* (New Delhi: Gitanjali Publishing House, 1982).

COMMISSION ON DEVELOPING COUNTRIES AND GLOBAL CHANGE, *For Earth's Sake* (Ottawa: International Development Research Centre, 1992).

CONYERS, D., 'Administration in China: some preliminary observations', *Journal of Administration Overseas*, vol. 16, 1977, pp. 98–113.

Development Dialogue, 'Education and Self-Reliance in Tanzania: a national perspective', vol. 2, 1978, pp. 37–50.

DUNSIRE, ANDREW, 'Bureaucratic Morality in the United Kingdom', *International Political Science Review*, vol.9, July 1988, pp. 179–191.

DUNSIRE, ANDREW, 'The Politics of Public Administration Reform', *The Changing Role of Government: Management of Social and Economic Activities*, Proceedings of a Commonwealth Roundtable held in London, 4–7 June 1991 (London: Commonwealth Secretariat, 1991) pp. 191–208.

DWIVEDI, O.P., 'Bureaucratic Corruption in Developing Countries', *Asian Survey*, vol. 7, no.4, April 1967, pp. 245–53.

DWIVEDI, O.P. 'India: Pollution Control Policy and Programmes', *International Review of Administrative Science*, vol. 43, no. 2, 1977, pp. 123–33.

DWIVEDI, O.P., *Public Service Ethics* (Brussels: International Institute of Administrative Sciences, 1978).

DWIVEDI, O.P., (ed.), *Environment and Resources: Policy Perspectives* (Toronto: McClelland & Stewart, 1980).

DWIVEDI, O.P. (ed.), *The Administrative State in Canada* (Toronto: University of Toronto Press, 1982).

DWIVEDI, O.P., 'Ethics and Administrative Accountability', *Indian Journal of Public Administration*, vol. 29, July–September 1983, pp. 504–17.

DWIVEDI, O.P., 'Ethics and Values of Public Responsibility and Accountability', *International Review of Administrative Science*, vol. 51, no.1, 1985, pp. 61–6.

DWIVEDI, O.P., (ed.), *Perspectives on Technology and Development* (New Delhi: Gitanjali Publishing, 1987).

DWIVEDI, O.P., 'Moral Dimensions of Statecraft: A Plea for an Administrative Theology', *Canadian Journal of Political Science*, vol. 20, no. 4, December 1987, pp. 699–707.

DWIVEDI, O.P., 'Ethics, the Public Service and Public Policy', *International Journal of Public Administration*, vol. 10, no. 1, 1987, pp. 21–50.

DWIVEDI, O.P., 'Man and Nature: A Holistic Approach to a Theory of Ecology', *The Environmental Professional*, vol. 10, no. 1, 1988, pp. 8–15.

DWIVEDI, O.P., 'Conclusion: A Comparative Analysis of Ethics, Public Policy, and the Public Service', in James S. Bowman and Frederick A. Elliston (eds), *Ethics, Government and Public Policy: A Reference Guide* (New York: Greenwood Press, 1988) pp. 307–22.

DWIVEDI, O.P., 'Administrative Heritage, Morality and Challenges in the Sub-Continent since the British Raj', *Public Administration and Development*, vol. 9, no. 3, July–August 1989, pp. 245–52.

DWIVEDI, O.P., 'Development Administration: Its Heritage, Culture and Challenges', *Canadian Public Administration*, vol. 33, no. 1, Spring 1990, pp. 91–8.

DWIVEDI, O.P., 'An Ethical Approach to Environmental Protection', *Canadian Public Administration*, vol. 35, no. 3, Autumn 1992, pp. 363–380.

DWIVEDI, O.P. and HENDERSON, KEITH M. (eds), *Public Administration in World Perspective* (Ames: Iowa State University Press, 1990).

DWIVEDI, O.P. and JAIN, R.B., *India's Administrative State* (New Delhi: Gitanjali Publishing House, 1985).

DWIVEDI, O.P. and JAIN, R.B. 'Administrative Culture and Bureaucratic Values in India', *Indian Journal of Public Administration*, vol. 36, no. 3, 1990, pp. 435–50.

DWIVEDI, O.P. and KISHORE, B., 'Protecting the Environment from Pollution: A Review of India's Legal and Institutional Mechanisms', *Asian Survey*, vol. 12, no. 9, September 1982, pp. 894–911.

DWIVEDI, O.P. and NEF, J., 'Crises and Continuities in Development Theory and Administration: First and Third World Perspectives', *Public Administration and Development* (UK), vol. 2, 1982, pp. 59–77.

DWIVEDI, O.P., NEF, J. and VANDERKOP, J., 'Science, Technology and Underdevelopment: A Critical Approach', *Canadian Journal of Development Studies*, vol. 11, no. 2, 1990, pp. 223–38.

DWIVEDI, O.P. and VENKATASAMY, V. (eds), *State of the Environment in Mauritius* (Port Louis: Ministry of Environment and Quality of Life, 1991).

ESMAN, MILTON D., 'The Politics of Development Administration', in John D. Montgomery and W.J. Siffin (eds), *Approaches to Development: Politics, Administration and Change* (New York: McGraw-Hill, 1966) pp. 69–70.

FARAZMAND, ALI (ed.), *Handbook of Comparative and Development Public Administration* (New York: Marcel Dekker, 1991).

FATHALY, O.I., PALMER, M. and CHACKERIAN, R., *Political Development and Bureaucracy in Libya* (Lexington: D.C. Heath, 1977).

FLORES, GILBERTO and NEF, JORGE (eds), *Administration Publica: Perspectivas Criticas* (San Jose: ICAP, 1984).

FONLON, BERNARD., *To Every African Freshman: the Nature, End and Purpose of University Studies* (Victoria: Camroon Times Press, 1969).

FREDERICKSON, H.G., 'Public Administration in the 1970s: Developments and Directions', *Public Administration Review*, vol. 26, no. 5, September–October 1976, pp. 564–76.

GEORGE, S., *How the Other Half Dies: The Real Reason for World Hunger* (New York: Penguin, 1976).

GIRVAN, N., BERNAL, R. and HUGHES, W., 'The IMF and the Third World: The Case of Jamaica, 1974–80', *Development Dialogue*, vol. 2, 1980, pp. 113–55.

GOULET, D., *The Cruel Choice: A New Concept in the Theory of Development* (New York: Atheneum, 1973).

GOULET, D., *The Uncertain Promise: Value Conflicts in Technology Transfer* (New York: IDOC/North America, 1977).

GOULET, D. and HUDSON, M., *The Myth of Aid* (New York: IDOC/North America, 1971).

HAQ, M. UL, 'An International Perspective on Basic Needs', *Finance and Development*, vol. 17, no. 3, 1980, pp. 11–14.

HEADY, FERREL, *Public Administration: A Comparative Perspective,* Fourth Edition (New York: Marcel Dekker, 1991).

HENDERSON, KEITH M., 'Rethinking the Comparative Experience: Indigenization versus Globalization' in O.P. Dwivedi and Keith M. Henderson (eds), *Public Administration in World Perspective* (Ames, Iowa: Iowa State University Press, 1990) pp. 333–41.

ISLAM, NASIR, 'Managing the Public Service under Structural Adjustment: Politics and Implementation of Administrative Reform', *The Changing Role of Government: Management of Social and Economic Activities*, Proceedings of a Commonwealth Roundtable held in London, 4–7 June 1991 (London: Commonwealth Secretariat, 1991) pp. 211–23.

JABBRA, JOSEPH G. and DWIVEDI, O.P. (eds.), *Public Service Accountability: A Comparative Perspective* (West Hartford, Conn.: Kumarian Press, 1988).

JONES, G.N., 'Frontiermen in Search of the "Lost Horizon"', *Public Administration Review*, vol. 36, no. 1, 1976, pp. 99–102.

KERNAGHAN, W.D.K. and DWIVEDI, O.P. (eds), *Ethics in the Public Service* (Brussels: International Institute of Administrative Sciences, 1983).

KIM, PAUL S., *Japan's Civil Service System* (New York.: Greenwood, 1988).

KOH, B.C., *Japan's Administrative Elite* (Berkeley: University of California Press, 1989).

KOOPERMAN, L. and ROSENBERG, S.,' The British Administrative Legacy in Kenya and Ghana', *International Review of Administrative Sciences*, vol. 43, no. 3, 1977, pp. 267–72.

LINDENBERG, MARC and CROSBY, BENJAMIN, *Managing Development: The Political Dimension* (New Brunswick: Kumarian Press, 1981).

MARINI, F., *Towards a New Public Administration: The Minnobrook Perspective* (Scranton: Chandler, 1971).

MEIER, G. (ed.), *Leading Issues in Economic Development* (New York: Oxford University Press, 1970)

MENDOZA, GABINO A., 'The Transferability of Western Management Concepts and Programs: An Asian Perspective', *Education and Training for Public Sector Management in Developing Countries* (New York: Rockefeller Foundation, 1977) pp. 61–71.

MONTGOMERY, JOHN D., *Bureaucrats and People, Grassroots Participation in Third World Development* (Baltimore, Md.: John Hopkins, 1988).

NAIDU, M.V., 'Ideology, Technology and Economic Development', in O.P. Dwivedi (ed.), *Perspectives on Technology & Development* (New Delhi: Gitanjali Publishing House, 1987) pp. 11–33.

NEF, J. and DWIVEDI, O.P., 'Development Theory and Administration: a fence around an empty lot?', *Indian Journal of Public Administration*, vol. 27, no. 1, January–March 1981, pp. 42–66.

NEF, J. and DWIVEDI, O.P., 'Training for Development Management: Reflections on Social Know-how as a Scientific and Technological System', *Public Administration and Development*, vol. 5, no. 3, 1985, pp. 235–49.

NGWA, COLLINS E.N., 'The Technological Variable in North–South Relations: An African Perspective' in O.P. Dwivedi (ed.), *Perspectives on Technology & Development* (New Delhi: Gitanjali Publishing, 1987) pp. 121–42.

NYERERE, J., *Freedom and Socialism* (Nairobi: Oxford University Press, 1970).

ORGANIZATION FOR ECONOMIC COOPERATION AND DEVELOPMENT, *The Internationalization of Software and Computer Services* (Paris: OECD, 1989).

PAUL, S., *Training for Public Administration and Management in Developing Countries* (Washington, DC: World Bank, 1983).

QADEER, M.A., 'External Precepts and Internal Views: Dialectic of Reciprocal Learning in Third World Urban Planning', in B. Sanyal (ed.), *Breaking the Boundaries* (New York: Plenum Press, 1991) pp. 193–210.

RAPHAELI, N. (ed.), *Readings in Comparative Public Administration* (Boston: Allyn and Bacon, 1967).

REDFORD, EMMETTE S., *Democracy in the Administrative State* (New York: Oxford University Press, 1969).

REINESMANN, HEINRICH, *New Technologies & Management: Training the Public Service for Information Management* (Brussels: International Institute of Administrative Sciences, 1987).

RIGGS, F., *Administration in Developing Countries* (Boston: Houghton -Mifflin, 1964).

ROMZEK, BARBARA S. and MELVIN J. DUBNICK, 'Accountability in the Public Sector: Lessons from the Challenger Tragedy', *Public Administration Review*, vol. 47, no. 3, May–June 1987, pp. 228–9.

ROSTOW, WALTER, *The Stages of Economic Growth: A Non-Communist Manifesto* (Cambridge, Mass.: Harvard University Press, 1962).

SADIK, NAFIS., 'Rethinking Modernism: Towards Human-Centred Development', *Development* (Journal of the Society for International Development) no. 2, 1991, pp. 15–20.

SANWAL, MUKUL (ed.), *Microcomputers in Development Administration* (New York: McGraw-Hill, 1987).

SCHAFFER, B., *The Administrative Factor* (London: Frank Cass, 1973).

SKOROV, G.E. (ed.), *Science, Technology and Economic Growth in Developing Countries* (New York: Maxwell House, 1979).

STENBERG, C.W., 'Contemporary public administration: challenge and change', *Public Administration Review*, vol. 36, no. 5, 1976, pp. 505–7.

STERN, BERNARD J., 'Some Aspects of Historical Materialism', in The Sponsoring Committee of the Bernard J. Stern Memorial Fund, *Historical Sociology: The Selected Papers of Bernard J. Stern* (New York: Citadel Press, 1959).

STONE, D.C., 'Tasks, Precedents and Approaches to Education for Development Administration', in the International Institute of Administrative Sciences' publication, *Education for Development Administration* (Brussels: Maison Ferdinand Larcier, 1966).

STOVER, WILLIAM J., *Information Technology in the Third World* (Boulder, Co.: Westview Press, 1984).

STREETEN, PAUL P., 'Problems in the Use and Transfer of an Intellectual Technology', in P.J. Lavakare, Ashok Parthasarathi and B.M. Udgaonkar (eds), *Scientific Cooperation for Development: Search for New Directions* (New Delhi: Vikas Publishing House, 1980) pp. 52–69.

SUBRAMANIAM, V. (ed.), *Public Administration in the Third World: An International Handbook* (Westport, CT: Greenwood, 1990).

SWERDLOW, I., *The Public Administration of Economic Development* (New York: Praeger, 1975).

SWAMINATHAN, M.S., 'The Environment Scene: Maladies and Remedies', *Survey of the Environment 1991* (published by *The Hindu*, Madras, India, 1991) pp. 4–7.

TAPIA-VIDELA, J., 'Understanding Organizations and Environments: a comparative perspective', *Public Administration Review*, vol. 36, no. 6, November–December. 1976, pp. 631–6.

THEOBALD, ROBIN, *Corruption, Development and Underdevelopment* (New York: Macmillan, 1990).

THOMAS, ROSAMUND, *The British Philosophy of Administration* (London: Longmans, 1978).

UK Report of the Fulton Committee, *The Civil Service, 1966–68*, vol. 1 (London: HMSO, 1968).

UNESCO, *Moving Towards Change* (Paris: UNESCO, 1976).

USA, Agency for International Development, *Cutting Edge Technologies and Microcomputer Applications for Developing Countries: Report of an Ad-Hoc Panel on the Use of Microcomputers for Developing Countries* (Boulder, Co.: Westview Press, 1989).

WALDO, DWIGHT, *The Administrative State* (New York: Ronald Press, 1948).

WALLERSTEIN, IMMANUEL, 'Crises: The World Economy, The Movements, and the Ideologies', in Albert Bergesen (ed.), *Crises in the World System*, vol. 6, Political Economy of the World-System Annals (Beverly Hills: Sage, 1983) pp. 21–36.

WALSH, ANNE MARIE, 'Public Administration and Development', in Institute of Public Administration, *IPA Report* (New York) vol. 1, no. 1, 1984, p. 10.

WORLD BANK, *World Development Report 1987* (New York: Oxford University Press, 1987).

WORLD BANK, *Economic Development with Environmental Management Strategies for Mauritius*, Report No. 7264–MAS (Washington DC: 1988).

WORLD BANK, *World Development Report 1991: The Challenge of Development* (New York: Oxford University Press, 1991).

WORLD BANK, *World Development Report 1992: Development and the Environment* (New York: Oxford University Press, 1992).

WORLD COMMISSION ON ENVIRONMENT AND DEVELOPMENT, *Our Common Future* (New York: Oxford University Press, 1987).

ZIPPER, RICHARD ISRAEL, *Un Mundo Cercano. El Impacto Politico y Econumico de las Nuevas Technoologias* (Santiago: Instituto de Ciencia Politica, Universidad de Chile, 1984).

Index